PLASTIC METHODS FOR STEEL AND CONCRETE STRUCTURES

Plastic Methods for Steel and Concrete Structures

Stuart S. J. Moy

Department of Civil Engineering,
University of Southampton

A HALSTED PRESS BOOK

JOHN WILEY & SONS
New York

First published in Great Britain 1981 by
The Macmillan Press Ltd

Published in the U.S.A. by
Halsted Press, a Division of
John Wiley & Sons, Inc.
New York

Printed in Hong Kong

Library of Congress Cataloging in Publication Data
Moy, Stuart S. J.
Plastic methods for steel and concrete structures.

"A Halsted Press book."
Bibliography: p.
1. Plastic analysis (Theory of structures)
2. Building, Iron and steel. 3. Reinforced concrete construction. I. Title.
TA652.M62 624.1′821 81-552
ISBN 0-470-27079-9 AACR2

CONTENTS

PREFACE

This book owes much to the many students, undergraduate and postgraduate, who have patiently sat through my lectures on the plastic methods. Their questions during and after the lectures have shown that certain aspects of the subject consistently create difficulties for them. I perhaps flatter myself that this is not due to shortcomings in my teaching technique, but because those points need greater attention than is possible in a lecture. It has been my intention to give this extra attention in the book.

The book is intended to cover most of the requirements of students at both undergraduate and postgraduate levels, but will also be of general interest to practising engineers who wish to use the plastic methods. The emphasis throughout is on the ideas behind, and application of, the plastic methods, rather than their mathematical justification (I have given references to appropriate texts for this). To do this I have simplified some of the arguments – I hope that this does not offend the purists, but I make no apology for it.

The book is divided into eight chapters. The first two deal with the concepts of plastic behaviour, and plastic bending in particular. Chapters 3 and 4 describe the various techniques for finding the collapse loads of steel frames. Chapter 5 shows how the plastic methods are used to design steel structures. Chapter 6 is divided into two parts. The first describes a method for calculating the deflections in a structure at the point of collapse. The second part deals briefly with the effect that those deflections, and instability, have on the collapse load of a structure. The effect is complex and detailed study is beyond the scope of this book, but it would be wrong to omit all reference to instability. Chapters 7 and 8 deal with reinforced concrete structures: chapter 7 examines the problems of applying the plastic methods to concrete frames, and chapter 8 describes the powerful yield-line and strip methods for slabs.

As a rough guide, based on my own courses, the first three chapters provide a solid introduction which would be appropriate to second-year undergraduate level, while chapters 4, 5 and parts of 6 and 8 would be suitable for a third-year course. A postgraduate, M.Sc., course could make use of all the material.

I have tried to be consistent in the layout of each chapter. There is an introduction which gives, where appropriate, the background theory. The meat of each chapter is usually presented in a series of examples which have been carefully graded. The first example introduces the ideas in the simplest possible manner, the subsequent ones introduce new ideas or examine those areas which can cause difficulties. The whole is then brought together briefly in a summary. At the end of each chapter (apart from the first) there is a series of examples, designed to bring out the various points made in the chapter. I cannot encourage the reader too strongly to work through the examples, preferably in order. Practice really is the best way of getting to grips with the plastic methods.

STUART S. J. MOY

NOTATION

It has been convenient for the same symbols to have different meanings or for similar things to have different symbols. These are all noted here, but are also defined in the text. It has been necessary to add suffixes to various symbols to indicate specific meanings – these are always defined in the text.

FORCES

C, T	compressive and tensile forces equivalent to bending stress blocks
F	axial member force in a pin-jointed truss
N	shear force in a bending member
P	axial force in a bending member
SF	shear force (abbreviation)
q	distributed loading on a slab
UDL	uniformly distributed load (abbreviation)
λ	load factor
λ'	change in load factor
λ_c	collapse load factor
V, H	vertical and horizontal applied loads or reactions
W	applied load
w, Q	line load

MOMENTS

BM	bending moment (abbreviation)
BMD	bending moment diagram (abbreviation)
FEM	fixed end moment (abbreviation)
M	moment; magnitude of a bending moment; moment of resistance per unit length
M_p	plastic moment
M_p'	reduced plastic moment
M_e	bending moment found by elastic analysis

M_n	moment per unit length normal to a yield-line
β_{red}	redistribution factor (CP 110 notation)

MATERIAL AND SECTION PROPERTIES

A	cross-sectional area
A_s	cross sectional area of steel reinforcement
b	width
d	over-all depth; effective depth of tensile reinforcement
d_1	effective depth of compression reinforcement
E	Young's modulus
E_{sh}	slope of stress–strain curve at the start of strain hardening
f_c	characteristic cube strength of concrete
G	structure weight function
G'	variable part of weight function
g	weight per unit length
I	second moment of area
l	effective length
l/r	slenderness ratio
r	radius of gyration $[=\sqrt{(I/A)}]$
S	plastic modulus
t_f	flange thickness
t_w	web thickness
x	depth to axis of zero strain in a concrete beam (CP 110 notation)
Z	section modulus
γ	shear strain
ϵ	direct strain
σ	direct stress
σ_y	yield stress
τ	shear stress
τ_y	yield shear stress

LENGTHS AND DISPLACEMENTS

e	eccentricity
L, h	overall dimensions
l, a, b	dimensions
R	radius
s	distance along a yield line
x, y, z	coordinate axes; distances in coordinate directions
$\alpha, \beta, \gamma, \theta, \phi$	angles; rotations
Δ, δ	displacements
χ	curvature

GENERAL

c, n, k	constants; ratios
i	ratio of positive and negative moments of resistance
K	Hillerborg correction factor
m	number of elementary mechanisms
n	number of plastic hinges
p	number of possible plastic hinges
r	degree of indeterminacy (redundancy)
YL	yield-line (abbreviation)
μ	measure of orthotropy in slabs

SIGN CONVENTION

The most convenient sign conventions have been used. Thus tensile stresses, strains and axial forces are assumed positive except in chapter 6 where compressive forces are assumed positive. Bending moments are plotted on the side of the member which is in tension. Tension on the underside of a beam or to the left of a column, as drawn, is assumed to be caused by a positive bending moment.

1 SOME GENERAL CONCEPTS

1.1 INTRODUCTION

The teaching of structural analysis follows, in general, traditional lines, the recognised route being statics, simple bending theory, virtual work and, finally, the analysis of rigid jointed structures. The main barrier that has to be crossed is from statically determinate structures (which can be analysed by statics alone) to statically indeterminate structures (which must, effectively, be analysed by a combination of statics and compatibility of deformations). The mathematics involved in the analysis of indeterminate structures frequently presents severe problems for students.

This 'classical' approach is based on the assumption that the stresses in the structure caused by the applied loads are within the elastic limit of the material used and thus deflections are small. The approach is, of course, widely used. However, an alternative has gained increasing support over the past 30 years or so. This new philosophy turns the problem on its head. It is obvious that any structure can be made to fall down (collapse) by applying loading of a sufficient magnitude. The purpose of the new analysis is to find that magnitude. It requires a knowledge of what happens at collapse and how structures behave when the stresses in the material exceed the elastic limit. This philosophy is embodied in the plastic methods of analysis and design. One important and reassuring feature of the plastic methods is that the mathematics involved is usually less formidable than with the traditional methods.

It is informative to examine the behaviour of structures from zero load to failure because it is then possible to show clearly the ideas behind the plastic methods. Two examples are presented to illustrate this behaviour, but before looking at the examples it is necessary to examine briefly the material used in the structure.

1.2 MILD STEEL – THE ALMOST PERFECT MATERIAL FOR PLASTIC ANALYSIS

The simplest mechanical test is to apply a controlled tensile force to a long bar

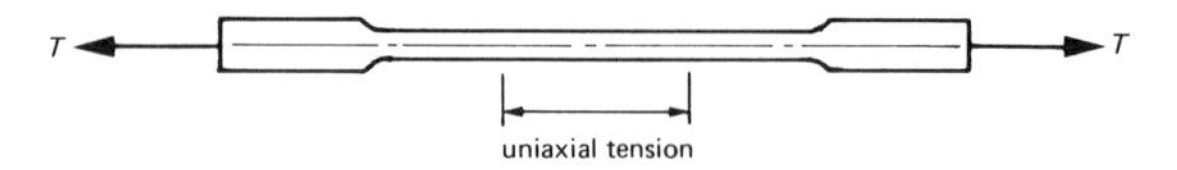

Figure 1.1

of the material (figure 1.1). In the middle of the bar, remote from the clamps at each end, a state of pure uniaxial tension exists.

If the extension of a mild steel specimen is measured (as strain) in this region and plotted against the applied force (expressed as a stress) the typical stress–strain curve shown in figure 1.2 is obtained. At small strains, stress is directly proportional to strain (region OU). The material is elastic, and the slope, E, is the *Young's modulus.*

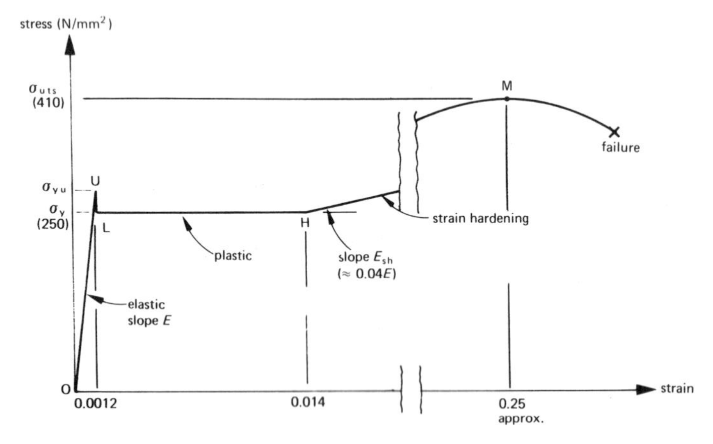

Figure 1.2

On average E is about 205 kN/mm^2. The point U is the limit of proportionality between stress and strain. When this limit is reached there is a rapid drop in stress to the point L. U is called the *upper yield point* with a corresponding stress σ_{yu}. The magnitude of σ_{yu} depends on the cross-sectional shape of the specimen and the type of equipment used to carry out the test. In many of the common structural steel sections which are hot rolled into shape, the residual stresses from the rolling process effectively remove point U. Hence the upper yield point is of no practical significance. The stress corresponding to point L is the *yield stress* σ_y with a typical magnitude for mild steel of 250 N/mm^2.

The strain at the yield stress is about 0.0012. When the strain is increased above this value it is found that no corresponding stress increase is required. The behaviour in the region LH of the graph is called *plastic* (increase in strain without change in stress is called *plastic flow*). The end of the plateau, H, is somewhat variable but a typical strain is 0.014. The strain in the plateau is thus at least ten times the strain at the yield point.

After H, an increase in strain requires an increase in stress, but the relation is now non-linear. This is called *strain hardening.* The initial slope E_{sh} of this region is about 4 per cent of Young's modulus, E. At a strain of at least 0.2, a 20 per cent increase in the length of the specimen, the stress reaches its maximum value (point M). This stress is called the *ultimate tensile strength* σ_{uts} and is about 410 N/mm^2. Further increase in strain produces necking and eventually a cup and cone fracture.

Careful tests have shown that the stress–strain curve for mild steel in compression is in fact identical to the one in tension up to the point of maximum stress, so that the complete graph is as in figure 1.3. If the specimen is loaded to, say, point X and the load then removed, initially the change in strain is elastic (slope E) as shown by the solid line XY. Ideal behaviour would follow the solid line with compressive plastic flow occurring when the stress reaches σ_y in the compressive sense. The actual behaviour follows the broken line XY′ indicating an apparently reduced yield stress in compression. The divergence from the ideal path is called the *Bauschinger effect.*

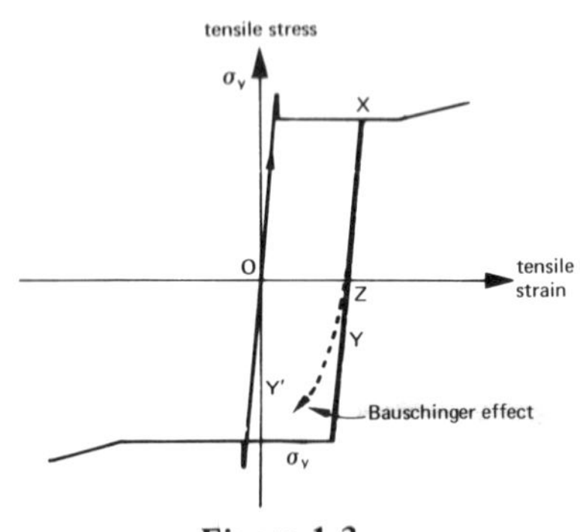

Figure 1.3

Perfect elastic–plastic behaviour is shown in figure 1.4. Mild steel can be made to fit this by

(1) ignoring the upper yield point. This causes no problems; many structural members do not show it anyway.

(2) ignoring strain hardening. This introduces some errors because many structures will have areas in the strain hardening region at collapse. However, the errors are small because of the small slope (E_{sh}) and are on the safe side since strain hardening represents an increase in strength.

(3) ignoring the Bauschinger effect. This causes errors but usually they are small. Figure 1.3 shows that when the stress is reduced to zero (point Z) there is little difference in the curves. In structures where full stress reversal is possible the errors can be significant.

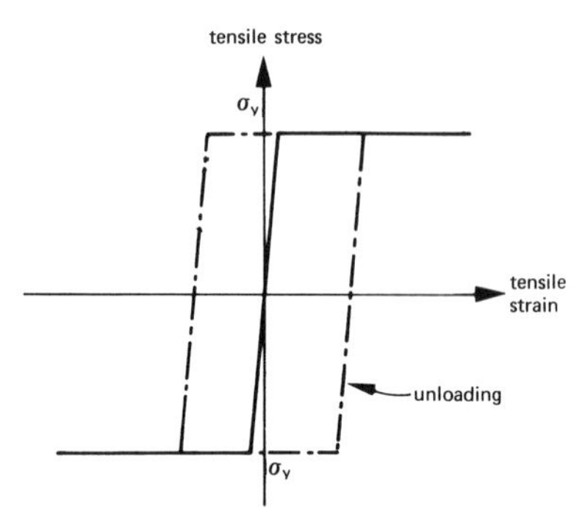

Figure 1.4

Mild steel is not the only structural steel – various higher strength grades are in common use. Higher strengths are achieved at the expense of ductility as shown in figure 1.5. In general, plastic analysis can be applied, with care, to structures made from these steels. This will be discussed further in chapter 6.

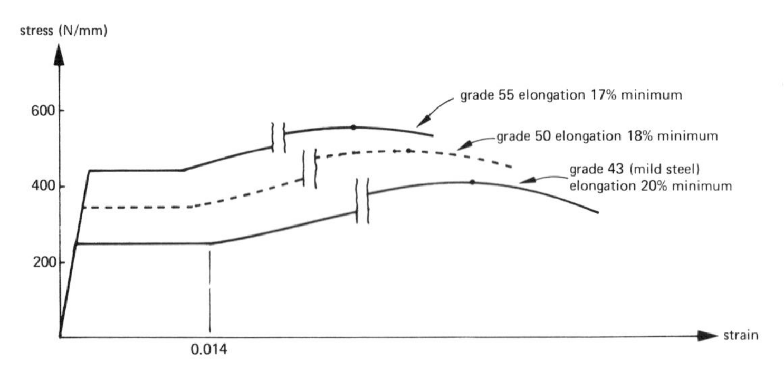

Figure 1.5

1.3 HOW STRUCTURES BEHAVE UNDER VARYING LOAD

Two examples of pin-jointed trusses are used to introduce several important ideas without too much conceptual or mathematical difficulty. The trusses will never be constructed for any practical use, but it will be convenient to imagine that they are made from mild steel.

1.3.1 Statically Determinate Tension Truss

The truss shown in figure 1.6 is statically determinate, which means it can be fully analysed by statics alone. The applied vertical force W at point O places both members OB and OC, in tension, the unknown internal forces being F_{OB} and F_{OC}.

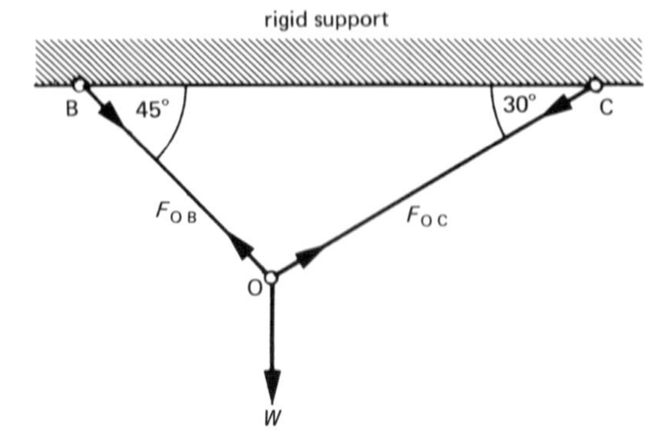

Figure 1.6

For horizontal equilibrium at O

$$F_{OB} \cos 45° = F_{OC} \cos 30°$$

that is

$$F_{OB} \frac{1}{\sqrt{2}} = F_{OC} \frac{\sqrt{3}}{2}$$

$$F_{OB} = 1.225 F_{OC} \tag{1.1}$$

Vertical equilibrium at O requires

$$F_{OB} \sin 45° + F_{OC} \sin 30° = W \tag{1.2}$$

Substitution of equation 1.1 into equation 1.2 gives the solution

$$\left.\begin{aligned} F_{OB} &= 0.897W \\ F_{OC} &= 0.732W \end{aligned}\right\} \tag{1.3}$$

Assume further that OB has a cross-sectional area of A and OC one of $2A$, so that

$$\left.\begin{aligned} \text{stress in OB} &= 0.897 \frac{W}{A} \\ \text{stress in OC} &= 0.366 \frac{W}{A} \end{aligned}\right\} \tag{1.4}$$

These stresses correspond to the two points shown on the stress–strain curve in figure 1.7. As W is increased the two points move up the line until the stress in member OB reaches the yield point. It is possible to find the load at which that occurs from

$$0.897 \frac{W}{A} = \sigma_y \tag{1.5}$$

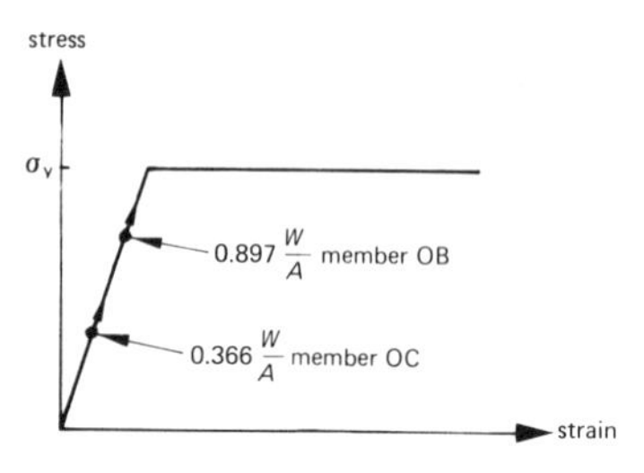

Figure 1.7

Any attempt to increase W results in plastic flow in OB with the stress fixed at σ_y. Mathematically equation 1.4, and hence equation 1.3, can no longer hold true because the stress in OB is now independent of W. Since equation 1.3 was found from the equilibrium equations 1.1 and 1.2, they also cannot be true and the structure is not in equilibrium. Physically OB increases in length without restraint and point O keeps moving. This loss of equilibrium is a definition of *collapse* of the structure. The greatest load at which equilibrium can be maintained is the *collapse load* (W_c). From equation 1.5

$$W_c = 1.115A\sigma_y \tag{1.6}$$

Two important points are illustrated by this example.

(1) In a statically determinate structure collapse occurs when the most heavily stressed member yields.

(2) The collapse load is directly proportional to the force which causes yield in that member ($A\sigma_y$). The constant (1.115 in this case) depends *only on the geometry* of the members.

1.3.2 Indeterminate Tension Truss

1.3.2.1 Analysis of Behaviour up to Collapse

Figure 1.8 shows a truss with three members and a vertical applied load W. There

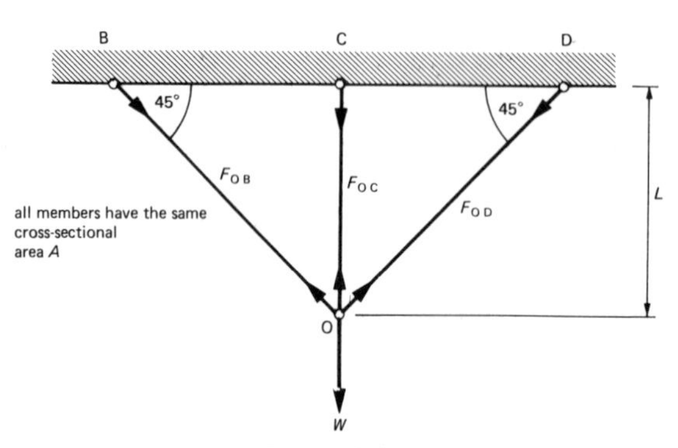

Figure 1.8

are three unknown member forces, so three independent equations are required to find those forces. Ghali and Neville [1] give a simple test for the indeterminacy of this type of truss, the test shows it to be one degree indeterminate. It will be necessary to consider both equilibrium and compatibility of deformations to find the unknowns.

For vertical equilibrium at O

$$F_{OB} \sin 45^\circ + F_{OC} + F_{OD} \sin 45^\circ = W \tag{1.7}$$

It is obvious from the symmetry, of the structure and loading, that O must move vertically downwards and OB must stretch the same amount as OD. Since the members are identical (in cross-section) this must imply that

$$F_{OB} = F_{OD} \tag{1.8}$$

Equations 1.7 and 1.8 cannot be solved to find the unknown member forces.

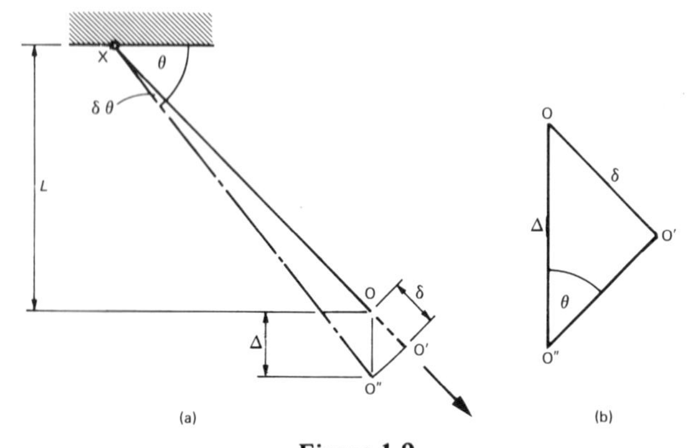

Figure 1.9

Consider now the sloping member shown in figure 1.9a. It extends by a distance δ when it carries a tensile force F_{OX}. In order for O to move vertically downwards to O″ the member (now extended to O′X) must rotate about X. If the extension is small compared to the length (the extension when the member yields is only about 0.1 per cent of the length) the angle $\delta\theta$ is negligibly small (if θ is 45°, $\delta\theta$ is about 4′ at the yield point). In triangle OO′O″ the angle OO″O′ is then equal to θ (see figure 1.9b) and

$$\delta = \Delta \sin \theta \tag{1.9}$$

If OX is elastic, stress/strain equals Young's modulus so that

$$\frac{F_{OX}}{A} \bigg/ \frac{\delta}{OX} = E$$

Substituting for δ and OX gives

$$F_{OX} = \frac{AE}{L} \Delta \sin^2 \theta \tag{1.10}$$

Compatibility is ensured in the truss if point O is constrained to move vertically. Hence equations 1.9 and 1.10 can be used for the members in the truss.

$$\delta_{OB} = \delta_{OD} = \Delta \sin 45° = \Delta/\sqrt{2}$$
$$\delta_{OC} = \Delta \sin 90° = \Delta \qquad (1.11)$$

and

$$F_{OB} = F_{OD} = \frac{AE}{L}\Delta\left(\frac{1}{\sqrt{2}}\right)^2 = \frac{AE\Delta}{2L}$$
$$F_{OC} = \frac{AE\Delta}{L} \qquad (1.12)$$

where Δ is vertical displacement of O. It can been seen from equation 1.12 that

$$2F_{OB} = F_{OC} \qquad (1.13)$$

Equation 1.13 together with equations 1.7 and 1.8 are the equations required to give the member forces. Substitute equation 1.8 into equation 1.7 to give

$$\sqrt{2}\, F_{OB} + F_{OC} = W \qquad (1.14)$$

then substitute equation 1.13 to give

$$\left(\frac{\sqrt{2}}{2} + 1\right) F_{OC} = W \qquad (1.15)$$

$$F_{OC} = 0.585W$$
$$F_{OB} = F_{OD} = 0.293W \qquad (1.16)$$

Yield starts in OC when the load is increased to W_1, given by

$$\frac{0.585W_1}{A} = \sigma_y$$

$$W_1 = 1.709\ A\sigma_y \qquad (1.17)$$

The stresses in all the members are shown in figure 1.10. OB and OD are stressed to $0.5\sigma_y$. The deflection of O when yield starts is found from equation 1.12

$$F_{OC} = A\sigma_y = \frac{AE\Delta_1}{L}$$

$$\Delta_1 = \frac{L\sigma_y}{E} \qquad (1.18)$$

Up to this stage the example has been a more complicated version of the example in section 1.3.1, but from now on there are important differences. Although OC has yielded, so that the force in it is limited to $A\sigma_y$, point O cannot move

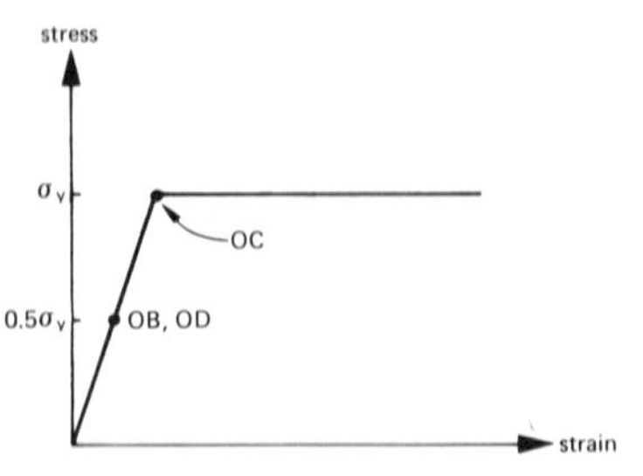

Figure 1.10

freely because it is restrained by the remaining elastic members. When W is increased there are only two unknown member forces, so OC yielding has caused a reduction (of one) in the degree of redundancy.† Equation 1.15 becomes

$$\surd(2)\, F_{OB} = W - A\sigma_y \tag{1.19}$$

so that

$$F_{OB} = 0.707\,(W - A\sigma_y) \tag{1.20}$$

The solution is obtained from the equilibrium equation without recourse to the compatibility equations, because the truss has become statically determinate. OB and OD yield when

$$F_{OB} = A\sigma_y \tag{1.21}$$

Substituting equation 1.20 into equation 1.21 gives the load W_2 when this occurs

$$0.707\,(W_2 - A\sigma_y) = A\sigma y$$

$$W_2 = 2.414\, A\sigma_y \tag{1.22}$$

Equation 1.12 (for member OB, which has remained elastic up to this load) gives the corresponding deflection Δ_2 of point O

$$\Delta_2 = \frac{2L\sigma_y}{E} \tag{1.23}$$

The member stresses at W_2 are plotted in figure 1.11. Since all the members are

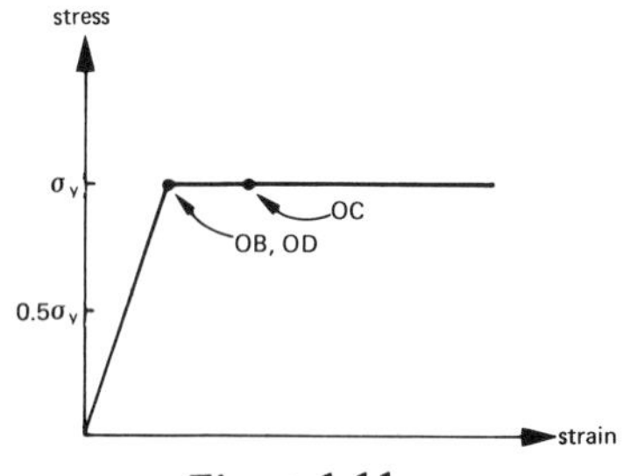

Figure 1.11

† Indeterminacy and redundancy are synonymous.

now yielding, W_2 is collapse load (W_c) of the truss. At the instant when W_2 is applied, called the *point of collapse*, the structure is in a state of unstable equilibrium. Any disturbance would destroy the equilibrium, plastic flow would occur in all the members, and point O would move without check.

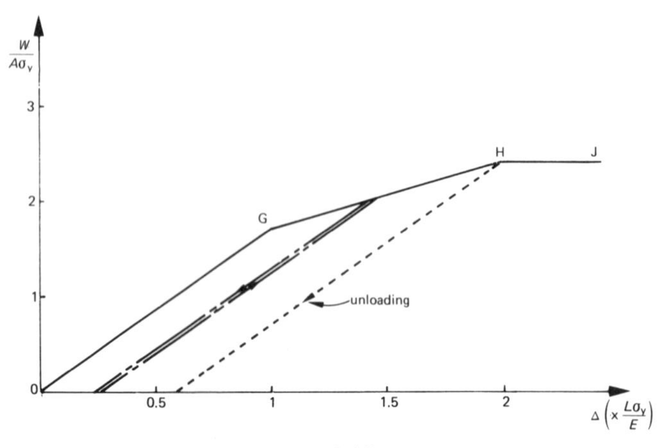

Figure 1.12

The results of the analysis in this section are summarised in figure 1.12 and tables 1.1 and 1.2. There are several important points which arise from the analysis, as follows.

(1) *The degree of indeterminacy (redundancy) of the structure is reduced by one each time a member yields.* This was noted in the analysis when member OC yielded When the truss became statically determinate one more member yielding would have caused failure (as in the example in section 1.3.1). In this case because of the symmetry two members yielded simultaneously. This reduction in redundancy provides a useful check. Provided the degree of redundancy of the structure can be found, it is a simple matter to determine the minimum number of yielding members required at collapse.

(2) *Redistribution of internal forces.* Table 1.1 shows the relative values of the internal forces at the key loads W_1 and W_c. It is clear that the relative

Table 1.1 Relative values of member forces

Load	W_1	W_c
F_{OC}/F_{OC}	1	1
F_{OB}/F_{OC}	0.5	1
$F_{OB} : F_{OC}$	1 : 2	1 : 1

magnitudes change drastically as yield starts in a member. The least heavily loaded members initially, gradually carry a larger proportion of the applied load. This is called redistribution of the internal forces and is an important characteristic of structural behaviour.

(3) *Reduction in stiffness.* Figure 1.12 is a graph of applied load W (plotted as $W/A\sigma_y$) against deflection Δ. The graph is characterised by three straight lines which gradually approach the horizontal. This is confirmed by the numerical values given in table 1.2. The slope (rate of change of load with respect to deflection, $dW/d\Delta$) is a measure of the stiffness of the structure. Initially, the stiffness is greatest, but each time a member yields (and there is a reduction in the degree of redundancy) there is a sudden drop in stiffness. In the final stage (HJ) when all the members are yielding the line is horizontal indicating zero stiffness. Hence an alternative *definition of collapse is when the stiffness of the structure becomes zero!*

Table 1.2

Region of figure 1.13	Slope $\frac{dW}{d\Delta}$ $\left(\times\frac{AE}{L}\right)$
OG	1.707
GH	0.707
HJ	0

1.3.2.2 What Happens when the Load is Removed after Yield has Occurred

It was shown earlier (section 1.2) that when a tension specimen is unloaded after yield, the stress reduction occurs elastically. To illustrate how this works in a redundant structure, consider the previous example at the instant when W_c has been applied, but before the last two members have undergone any plastic flow. If an additional load of $-W_c$ is applied the net effect is zero load. Since unloading occurs elastically in every member, equations 1.16 will give the *changes* in member forces due to $-W_c$

$$\begin{aligned}\text{change in } F_{OC} &= -0.585\ W_c\\ \text{change in } F_{OB} &= -0.293\ W_c\end{aligned} \tag{1.24}$$

Substituting for W_c from equation 1.22, the resultant forces in the members at zero load are

$$F_{OC} = A\sigma_y - 0.585 \times 2.414\, A\sigma_y = -0.414 A\sigma_y$$
$$F_{OB} = A\sigma_y - 0.293 \times 2.414\, A\sigma_y = 0.293 A\sigma_y \qquad (1.25)$$

Equations 1.25 show that at zero load the forces in the members have not returned to zero. These 'residual forces' have an important property which can be shown by substituting equations 1.25 into the equilibrium equation 1.15

$$\sqrt{2} \times 0.293\, A\sigma_y - 0.414 A\sigma_y = 0$$
$$(0.414 - 0.414)\, A\sigma_y = 0$$
$$0 = 0$$

In other words the residual forces are in equilibrium with each other. They are called a set of *self-equilibrating residual forces.* A similar analysis gives the residual deflection Δ_r at zero load

$$\Delta_r = \frac{2L\sigma_y}{E} - \frac{0.585 \times 2.414\, A\sigma_y L}{AE} = \frac{0.588 L\sigma_y}{E} \qquad (1.26)$$

The unloading process is shown by the broken line in figure 1.12.

Notice that in this example the residual forces are all smaller than the load ($\pm A\sigma_y$) at which yield would occur. The example was chosen deliberately to ensure this but it is apparent that in some structures the redistribution of internal forces could be such as to cause yield in the opposite sense on unloading. Of course, the mathematics involved in analysing that situation are rather more difficult and there is the additional problem of the Bauschinger effect.

If on reaching zero load the truss is reloaded in the same manner the behaviour follows the 'unloading' line in the reverse direction. This is as expected because all members start off elastic and the increase in load will cause elastic changes until yield occurs. Yield now occurs simultaneously in all three members when the load W_c is reached.

It can be concluded, therefore, that unloading after any member has yielded alters the way in which the member forces change, but has no effect on the collapse load. This argument applies if the unloading is made before the point of collapse is reached, as is also shown in figure 1.12.

1.3.2.3 What Happens if the Members are not Manufactured to the Correct Length

When some of the members in an indeterminate structure are the wrong length – a situation which is often called 'lack of fit' – it is necessary to force the joints together. This produces a system of forces in the members. To illustrate what happens, consider the same structure as before, but with member OC too short by $0.75L\sigma_y/E$, as in figure 1.13. In order to close the gap, it is necessary to apply the forces marked F. In the sloping members, figure 1.13a. F pushes O up by a distance δ_1. In the other member OC, figure 1.13b, F pulls O down by δ_2. In

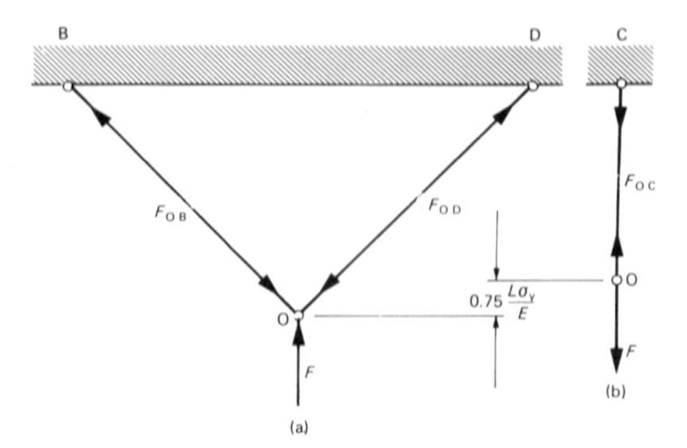

Figure 1.13

figure 1.13, due to the symmetry

$$F_{OB} = F_{OD}$$

For vertical equilibrium, figure 1.13a (compressive member force is negative)

$$\sqrt{(2)}F_{OB} = -F$$

For vertical equilibrium, figure 1.13b

$$F_{OC} = F$$

Substituting in equation 1.12 (downwards deflection is positive) gives

$$\delta_1 = \frac{2F_{OB}L}{AE} = \frac{-\sqrt{(2)}FL}{AE}$$

$$\delta_2 = \frac{F_{OC}L}{AE} = \frac{FL}{AE} \tag{1.27}$$

In order to close the gap and connect the members

$$|\delta_1| + |\delta_2| = \frac{0.75L\sigma_y}{E} \tag{1.28}$$

Substitute equation 1.27 into equation 1.28 to give

$$(\sqrt{2} + 1)\ \frac{FL}{AE} = \frac{0.75L\sigma_y}{E}$$

so that

$$F = 0.311A\sigma_y$$

$$F_{OB} = -0.220A\sigma_y$$

$$F_{OC} = 0.311A\sigma_y \tag{1.29}$$

$$\delta_1 = \frac{-0.439L\sigma_y}{E}$$

In the process of joining the members the forces F cancel each other to give zero applied load on the structure. Thus the initial effect of members of incorrect length is to introduce a set of self-equilibrating residual forces and permanent deformations, similar to those discussed in the previous section.

When the load W is hung on the truss, the member forces change. The behaviour of the truss as W increases can be followed in the same manner as in section 1.3.2.1, and is summarised in table 1.3. Load deflection curves for this

Table 1.3

$W/A\sigma_y$	$F_{OB}/A\sigma_y$	$F_{OC}/A\sigma_y$	$\Delta\,(\times\frac{L\sigma_y}{E})$	Comments
0	−0.220	0.311	−0.439	
1.178	0.125	1	0.250	OC yields
2.414	1	1	2	collapse

case and the perfect truss are plotted in figure 1.14. At lower loads the curves are markedly different but as yield spreads to the various members they come together. At the point of collapse the load and deflection are identical.

'Lack of fit' in the members has no effect on the load at which collapse occurs because of redistribution of the internal forces. The redistribution may occur in a different sequence and require additional plastic flow when there is 'lack of fit' but it gives the same result.

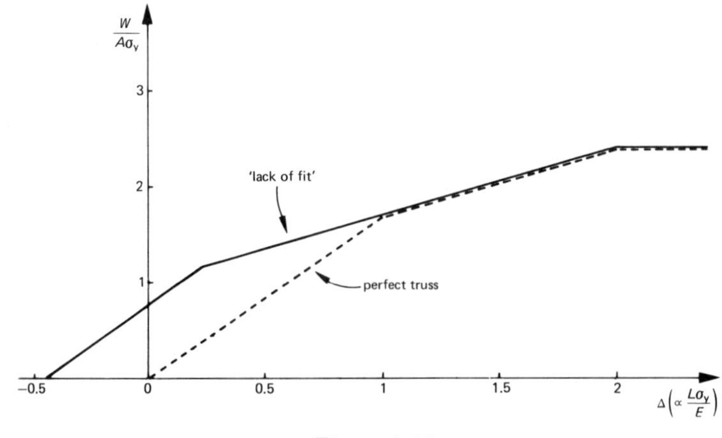

Figure 1.14

1.3.2.4 Calculation of the Collapse Load Alone makes Life much Easier

Since there is no need to worry about lack of fit of the individual members when considering collapse, the calculation of the collapse load is very easy. At collapse the force in each member is $A\sigma_y$, thus for vertical equilibrium at O

$$2A\sigma_y \sin 45^\circ + A\sigma_y = W_c$$

$$W_c = 2.414A\sigma_y$$

Compare the simplicity of this calculation with the elastic analysis in section 1.3.2.1. Plastic analysis produces this relative simplicity even in more complicated structures.

1.4 SUMMARY

The purpose of this chapter has been to introduce various important *ideas* and get the reader to *think* about the behaviour of structures. The illustrative examples were carefully chosen for that purpose. In fact, plastic analysis is not used for trusses because in a practical situation some truss members will be in compression, and hence liable to buckle. Plastic analysis has found its widest application in structures which carry load by bending. However, the ideas that have been illustrated by the examples in this chapter are common to all structures, so it is worth summarising them before looking more closely at structures with members in bending.

(1) The calculation of the collapse load is not difficult. The mathematics is far simpler than that required to find the load at first yield.

(2) The redundancy of the structure is reduced as members yield. One member yielding causes the loss of one degree of redundancy. At the point of collapse the structure is statically determinate.

(3) There is a reduction in stiffness as members yield. At collapse the stiffness is reduced to zero.

(4) The point of collapse is a state of unstable equilibrium. Once this is disturbed the structure collapses and equilibrium cannot be restored.

(5) There is redistribution of the internal forces as members yield.

(6) Lack of fit of the individual members has no effect on the collapse load.

2 PLASTIC BENDING

2.1 INTRODUCTION

In the previous chapter the behaviour of trusses was discussed. However, a more common form of construction nowadays is the framed structure with joints capable of transmitting bending moments. It will become clear that the ideas already developed can be applied to such structures.

In the trusses the applied forces were transmitted to the supports by means of axial forces in the individual members. These axial forces were also described as the internal forces. In a framed structure the applied forces are resisted mainly by shear forces and bending moments in the members (there will be some axial forces as well, but their effect will normally be secondary, except possibly in the columns).

Before considering the collapse of frames it is necessary to examine what happens to a member when it bends enough to cause plastic flow in the most highly stressed parts.

2.2 WHAT HAPPENS TO A BEAM WHEN IT BENDS?

The simplest structure which resists load by bending is the simply supported beam, as shown in figure 2.1a. It has an effective span L and carries a central vertical load W. The load is carried to the supports by the beam bending. Equilibrium gives a vertical reaction of $W/2$ at each support. The support system

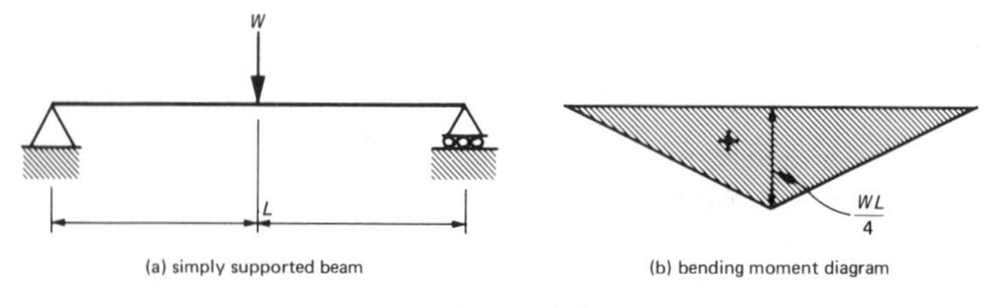

Figure 2.1

ensures no bending moments or horizontal reactions at the supports. The bending moment (BM) diagram for the beam is shown in figure 2.1b. The beam is sagging over its whole length with a maximum bending moment of $WL/4$ under the point load. It is the behaviour of the cross-section at the point of maximum BM that must now be examined in detail.

Simple bending theory (based on elastic material behaviour) gives the following information about the section. If there is no yield in the material there are straight line relationships for stress and strain over the whole depth of the section, as in figure 2.2. The level at which stress and strain are zero is the ***axis of zero strain***. The term 'axis of zero strain' has been used deliberately, to avoid confusion with the 'neutral axis' which is associated by common usage with the centroid. Stress and strain are proportional to distance (z) from this axis and for sagging there is maximum compression at the top and tension at the bottom. The maximum stress is given by

$$\sigma_{max} = \frac{M}{Z}$$

where M = bending moment and Z = section modulus (minimum). (Notice that for an asymmetric section, as in figure 2.2, bending about the y-axis, there are two possible values of the section modulus.

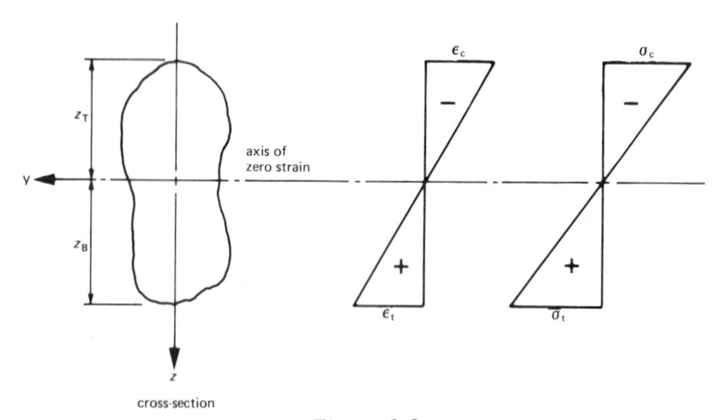

Figure 2.2

$Z_T = I/z_T$ and $Z_B = I/z_B$ where I = the second moment of area of the section about the y-axis. There will be different intensities of stress at the top and bottom).

There will be elastic behaviour until the maximum stress reaches the yield point. At this stage, of course, only material at the outside edge of the section is yielding. It has been shown in tests that the distribution of strain stays linear over the depth of the section after yield (and the simple bending theory assumption of plane sections remaining plane is still valid). It is possible to find the stress at any position from the stress–strain curve, as shown in figures 2.3

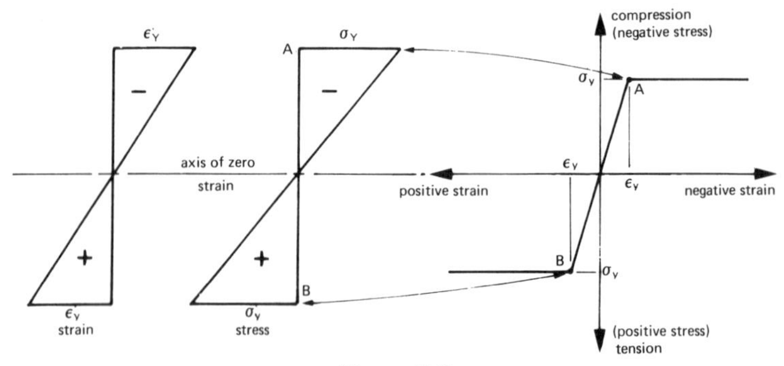

Figure 2.3

and 2.4. As the bending moment is increased, yielding spreads towards the axis of zero strain. The stress distribution shows two constant regions where yield has occurred (the stress is limited to the yield stress, but strain can increase by plastic flow), joined by a linear (elastic) stress distribution. The logical conclusion is shown in figure 2.5 with constant stress to the axis of zero strain. With all the

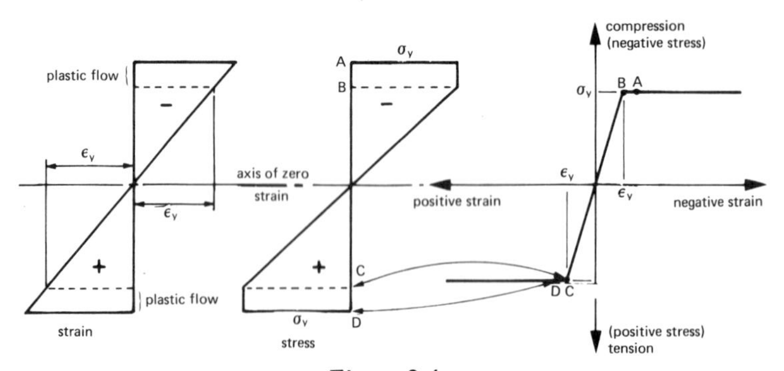

Figure 2.4

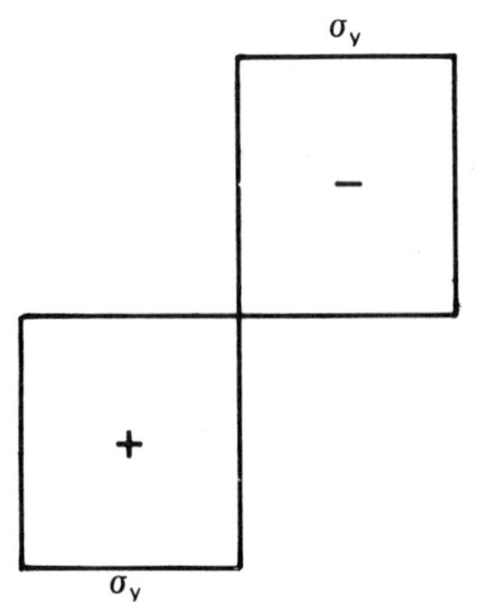

Figure 2.5

material yielding (in compression above and in tension below the axis of zero strain) the section behaves like a hinge because the strain can increase everywhere in the section without any change in stress. This hinge action is illustrated in figure 2.6. The section has become a *plastic hinge.* The plastic hinge is formed at a BM equal to the *plastic moment of resistance* of the section, which is the largest bending moment which the section can carry. It is usually shortened to *plastic moment* and given the symbol M_p.

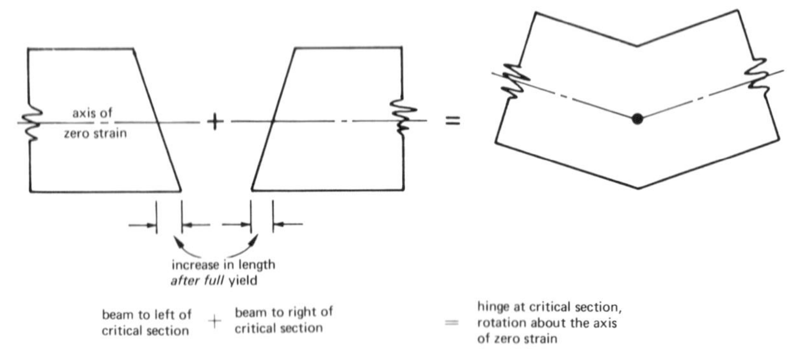

Figure 2.6

When the plastic hinge forms in the simply supported beam, collapse occurs (see figure 2.7). This is directly comparable to the statically determinate truss in section 1.3.1 where collapse occurred when the first member yielded. In this case the beam is determinate and collapses when one hinge forms. It is important to realise that the formation of a plastic hinge (not first yield) in a bending member is equivalent to yield in an axially loaded member. When collapse occurs, the load and right-hand support must move. The beam has become a *mechanism.* It is still easy to find the collapse load W_c, by equating the maximum moment due to the applied load to the plastic moment of the beam.

$$\frac{W_c L}{4} = M_p$$

that is

$$W_c = \frac{4M_p}{L} \tag{2.1}$$

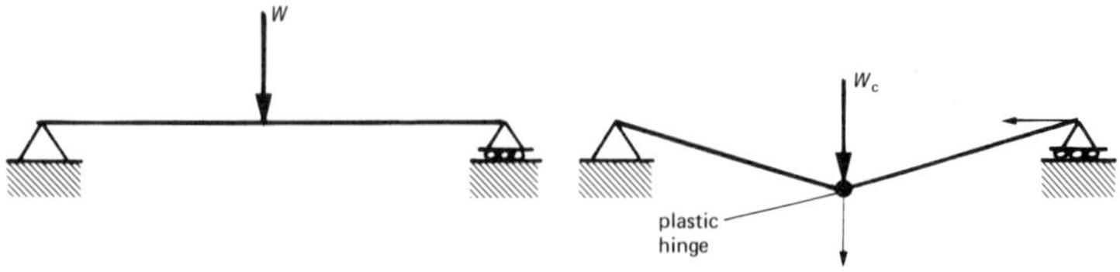

Figure 2.7

2.3 CALCULATION OF THE PLASTIC MOMENT

2.3.1 General

A general cross-section is shown in figure 2.8. The stress distribution due to the formation of a plastic hinge by bending about the y-axis is also shown. Since the

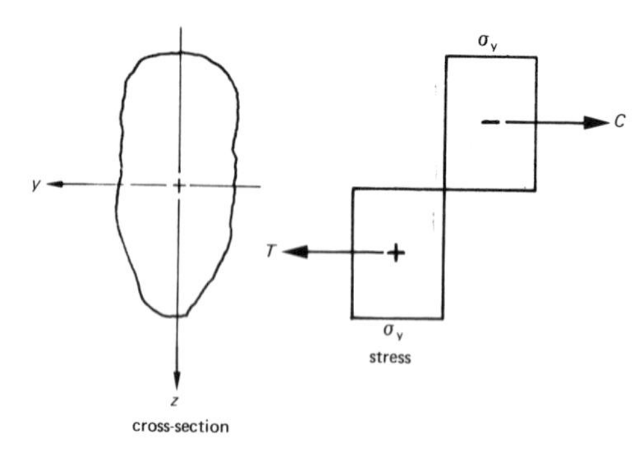

Figure 2.8

hinge has been formed by bending only, horizontal equilibrium of the section requires that

$$C = T \tag{2.2}$$

where C is the compressive force due to compressive yield above the axis of zero strain and T is the tensile force due to tensile yield below the axis. Hence

$$\text{area of section in compression} \times \sigma_y = \text{area of section in tension} \times \sigma_y \tag{2.3}$$

Equation 2.3 shows that the axis of zero strain, when a plastic hinge forms, bisects the cross-sectional area. This axis only coincides with the centroid of the section when the section is symmetric about the axis of zero strain.

2.3.2 Rectangular Section

In a rectangular section (figure 2.9) bending about the y-axis, the axis of zero strain is $d/2$ from the top of the section.

$$C = T = \frac{bd}{2}\sigma_y \tag{2.4}$$

Since these forces are caused by a bending moment equal to M_p taking moments

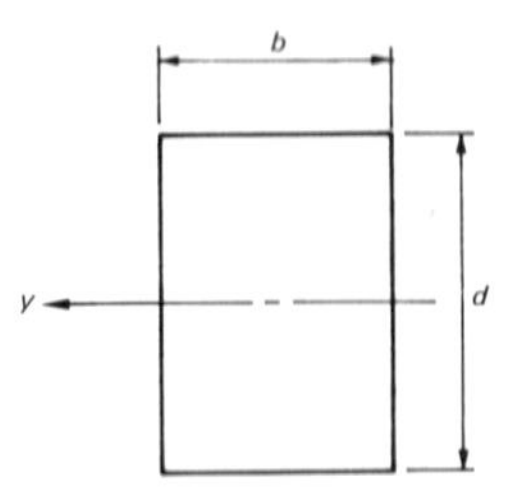

Figure 2.9

about the axis of zero strain gives

$$M_p = C \times \frac{d}{4} + T \times \frac{d}{4}$$

$$= 2 \times \frac{bd}{2}\, \sigma_y \times \frac{d}{4}$$

that is

$$M_p = \frac{bd^2}{4}\, \sigma_y \tag{2.5}$$

this is sometimes written

$$M_p = S\sigma_y \tag{2.6}$$

where S is called the ***plastic modulus*** of the section (cf. the section modulus Z). The plastic modulus is a geometric property of the cross-section. The ratio of plastic modulus to section modulus is the *shape factor* of the section

$$\text{shape factor} = \frac{S}{Z} \tag{2.7}$$

For the rectangular section $Z = bd^2/6$ so that

$$\text{shape factor} = \frac{bd^2}{4} \bigg/ \frac{bd^2}{6} = 1.5$$

The approach given for the rectangular section can be generalised to more complex shapes, as follows.

(1) Find the axis that bisects the cross-sectional area.

(2) Divide the cross-section into simple shapes whose properties can be found easily. For structural sections these will be rectangles.

(3) The plastic modulus is the sum of the plastic moduli of each rectangle.

The only complication in (3) is that the centroid of the rectangles may not coincide with the axis of zero strain of the whole section, as shown in figure 2.10.

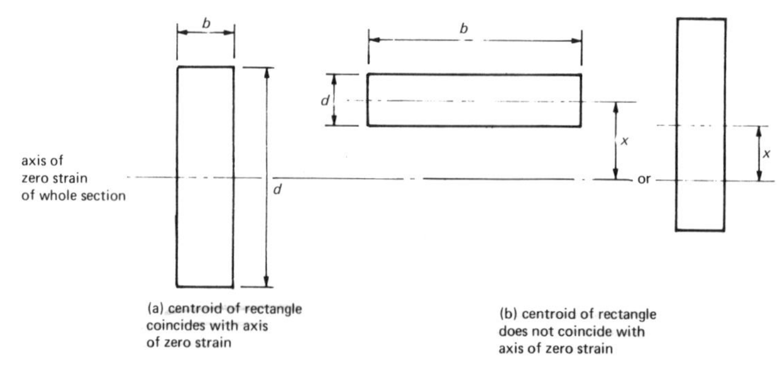

Figure 2.10

When the centroid coincides with the axis of zero strain, equation 2.5 can be used to give

$$\text{plastic modulus of rectangle} = \frac{bd^2}{4}$$

When the centroid does not coincide, a similar approach to the one above shows that

$$\text{plastic modulus of rectangle} = bdx \tag{2.8}$$

2.3.3 I-section

This is probably the most common structural section. A typical shape is shown in figure 2.11a. The shape is produced by passing a billet of steel through a series

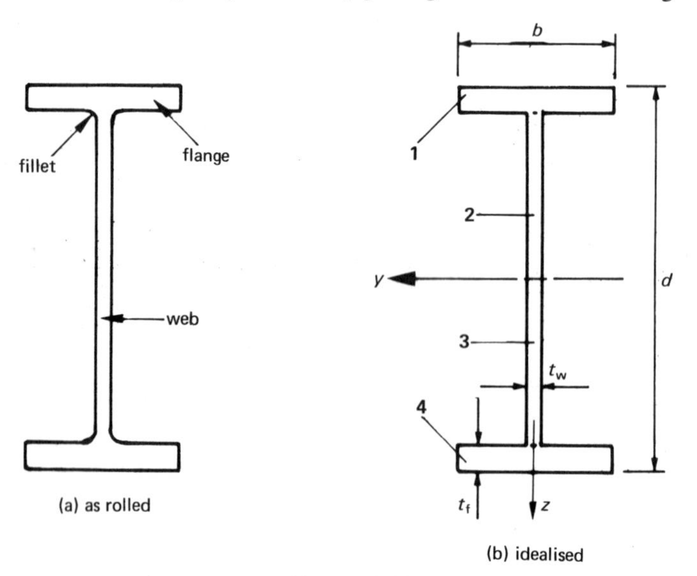

Figure 2.11

of rollers. Modern techniques produce sections with parallel faces to both the flanges and web. An I-section is symmetric about the y-and z-axes. In calculations the small fillets between the web and flanges are ignored. This idealised section is shown in figure 2.11b, with relevant dimensions and a convenient division into four rectangles. Table 2.1 shows how the plastic modulus of an I-section can be found using the method given in the previous section.

Table 2.1

Step	Bending about y-axis	Bending about z-axis
(1) axis of zero strain	y-axis	z-axis
(2) division into simple shapes	4 rectangles as in figure 2.11	4 rectangles as in figure 2.11
(3) plastic modulus of each area		
1	$bt_f\left(\frac{d}{2}-\frac{t_f}{2}\right)$	$\frac{b^2 t_f}{4}$
2	$\frac{(d-2t_f)}{2} \times t_w \times \frac{(d-2t_f)}{4}$	$\frac{(d-2t_f)}{2} \frac{t_w{}^2}{4}$
3	$\frac{(d-2t_f)}{2} \times t_w \times \frac{(d-2t_f)}{4}$	$\frac{(d-2t_f)}{2} \frac{t_w{}^2}{4}$
4	$bt_f\left(\frac{d}{2}-\frac{t_f}{2}\right)$	$\frac{b^2 t_f}{4}$
(4) plastic modulus	$bt_f\,(d-t_f)+\frac{t_w(d-2t_f)^2}{4}$	$\frac{b^2 t_f}{2}+\frac{(d-2t_f)t_w{}^2}{4}$

A typical section is a 457 x 152 UB 82 (the shorthand means a universal beam, with nominal depth 457 mm, nominal width 152 mm and weighing 82 kg/m). The Constrado handbook [2] gives the actual dimensions of the section as d = 465.1 mm, b = 153.5 mm, t_w = 10.7 mm and t_f = 18.9 mm. These give plastic modulus values of 1.783×10^6 mm^3 and 0.235×10^6 mm^3 for bending about the y- and z-axes respectively.

In fact it is possible to reduce the amount of effort in table 2.1 by making use of the symmetry of the section. Notice that the plastic moduli for **1** and **4**, and **2** and **3** are identical. Thus all that is necessary is to find the values for **1** and **2** and double the result.

Typical shape factors for I-sections are 1.15 for bending about the y-axis and 1.6 for bending about the z-axis. The actual values depend on the dimensions of the section.

2.3.4 Asymmetric Sections

Asymmetric sections do not yield simultaneously at the top and bottom of the section. As a result the axis of zero strain moves from the centroid before yield, to the axis which bisects the cross-sectional area.

To illustrate this, section modulus and plastic modulus calculations are shown for bending of a T-section (figure 2.12) about the y-axis in table 2.2. The calculations show that the zero strain axis moves upwards as plastic flow spreads through the section. The shape factor is 1.817, which is typical for this type of section.

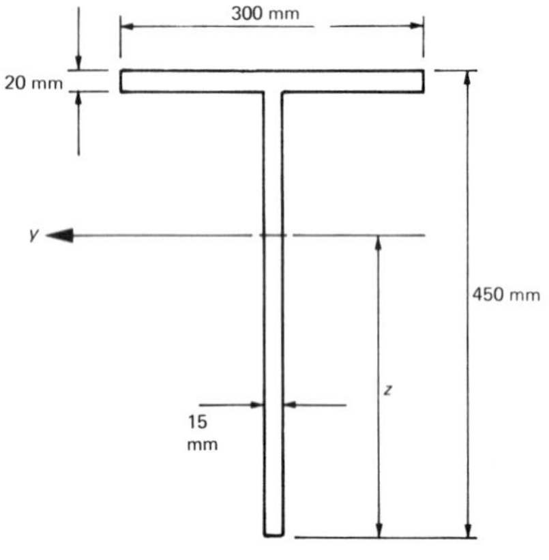

Figure 2.12

2.4 WHY THE PLASTIC MOMENT AND PLASTIC HINGE ARE IDEALISATIONS

It is possible to analyse the spread of yield through a section. In the trusses in chapter 1 it was done by reference to the material stress–strain curve. For members in bending it must be done by finding the relationship between bending moment and curvature for the section.

Assume that a short length of beam, length δx, which is initially straight, is bent into an arc of a circle as in figure 2.13. (This assumption is only true when the bending moments along the beam are constant, but the error is small when the moments vary provided that the deflection of the beam is small.) It has already been assumed that plane sections remain plane and the distribution of strain across the depth of the section is always as in figure 2.3, whatever the

Table 2.2

Section modulus calculation	Plastic modulus calculation
(i) position of centroid	(i) position of equal area axis
$A = 300 \times 20 + 430 \times 15$	$15 \times z = 300 \times 20 + (430 - z) \times 15$
$= 12450 \text{ mm}^2$	$15z = 6000 + 6450 - 15z$
$\Sigma Az = 6000 \times 440 + 6450 \times 2.5$	$z = \dfrac{12450}{30} = 415 \text{ mm}$
$= 4026750 \text{ mm}^3$	
$\bar{z} = \dfrac{4\,026\,750}{12450} = 323.4 \text{ mm}$	
centroid = 323.4 mm from bottom	equal area axis = 415 mm from bottom
$I = \dfrac{300 \times 20^3}{12} + 6000 \times 116.6^2$	$S = 6000 \times 25 + \dfrac{15 \times 15^2}{2}$
$+ \dfrac{15 \times 430^3}{12} + 6450 \times 108.4^2$	$+ \dfrac{15 \times 415^2}{2} = 150\,000 + 1700 +$
$= 200\,000 + 81\,573\,000$	$1\,291\,700 = 1\,443\,400 \text{ mm}^3$
$+ 99\,384\,000 + 75\,791\,000$	$S = 1\,443\,400 \text{ mm}^3$
$= 256\,948\,000 \text{ mm}^4$	Shape factor $= \dfrac{1\,443\,400}{794\,500}$
$Z = \dfrac{256\,948\,000}{323.4} = 794\,500 \text{ mm}^3$	$= 1.817$

Note. The value of Z quoted gives the maximum stress in the section, which is at the bottom

distribution of stress across the section. The arc which defines the axis of zero strain must remain δx long, while

$$\text{length of top arc} = \delta x\,(1 - \epsilon_C)$$

$$\text{length of bottom arc} = \delta x\,(1 + \epsilon_T)$$

From the geometry of figure 2.13

$$R\theta = \delta x \tag{2.9}$$

$$(R + z_2)\theta = \delta x\,(1 + \epsilon_T) \tag{2.10}$$

$$(R - z_1)\theta = \delta x\,(1 - \epsilon_C) \tag{2.11}$$

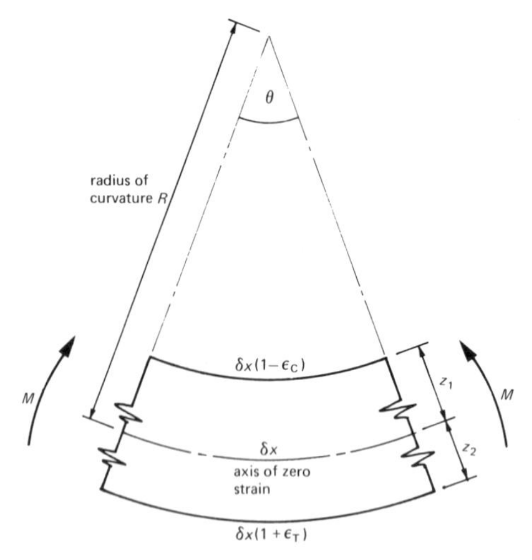

Figure 2.13

substituting equation 2.9 into equations 2.10 and 2.11 gives

$$R + z_2 = R(1 + \epsilon_T) \tag{2.12}$$

$$R - z_1 = R(1 - \epsilon_C) \tag{2.13}$$

subtracting equation 2.13 from equation 2.12 gives

$$z_1 + z_2 = R(\epsilon_T + \epsilon_C)$$

The inverse of the radius of curvature, R, is defined as the curvature, χ, thus

$$\text{curvature } \chi = \frac{\epsilon_T + \epsilon_C}{z_1 + z_2} = \frac{\text{range of strain}}{\text{depth of section}} \tag{2.14}$$

Curvature is simply a measure of bending deformation.

The 'ideal' elastic–plastic moment curvature relationship is shown in figure 2.14a. There is an elastic portion where an increase in curvature requires an increase in the moment causing the curvature. When the moment reaches the plastic moment, a plastic hinge forms and the curvature can increase without any change in the moment. This is plastic rotation of the hinge. From simple bending theory the slope of the elastic portion is EI.

It was assumed in section 2.2 that when a plastic hinge has formed, the whole of the critical section has yielded. This implies that just above and below the axis of zero strain there are strains equal to the yield strain, but of opposite sign. The only way that this and the assumption of plane sections remaining plane can be valid, is for the strains to be infinitely large at the top and bottom of the section. Clearly this is physically impossible.

The spread of yield through the section is analysed by assuming a distribution of stress as in figure 2.15, and varying the size of the elastic region. At each

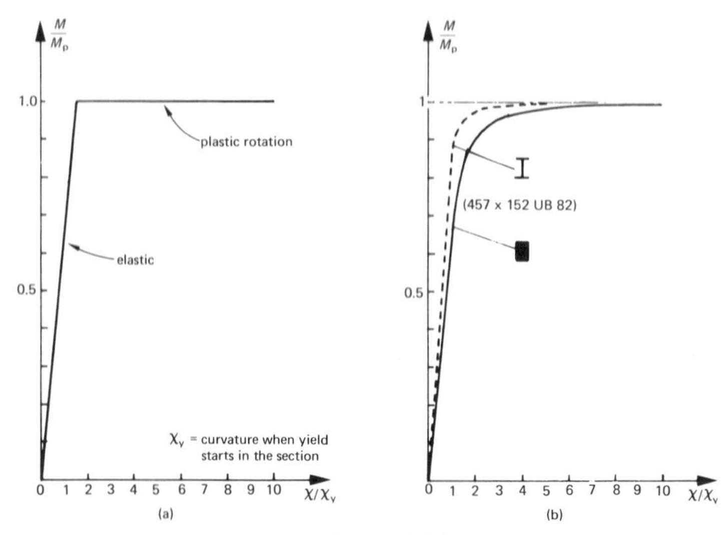

Figure 2.14

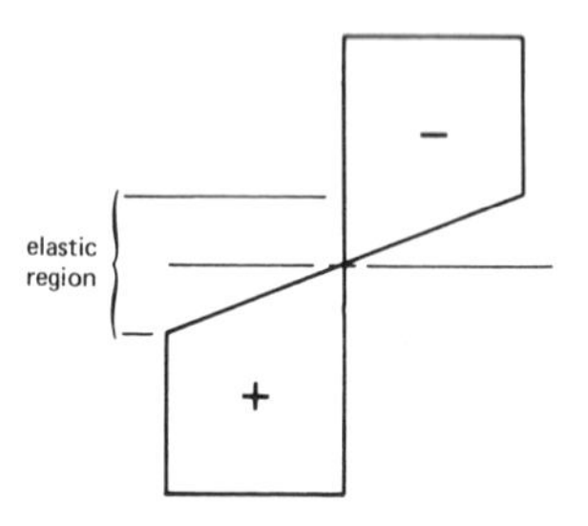

Figure 2.15

stage the moment which causes the stress distribution and the strains at the top and bottom of the section can be calculated. These can be used to plot the actual moment–curvature relationship of the section. This has been done in figure 2.14b for two sections, and as can be seen the actual curves are asymptotic to the ideal relationship. Practically, there is always a small elastic region sandwiched in the middle of the section. The assumption of ideal behaviour introduces very small errors for sections with shape factors close to unity (I-beam $\approx$ 1.15), whereas sections with higher shape factors are far from ideal.

So far only the critical section with highest bending moment has been considered, and it would appear that plastic flow is confined to that section. In fact, the bending moments at sections adjacent to the critical one are sufficient to cause yielding before the plastic hinge can form. The result is zones of plastic material around the critical section, as shown in figure 2.16. The extent of the zones depends on the type of loading. It is greater with distributed loads where the changes in bending moment are more gradual, than with concentrated loads.

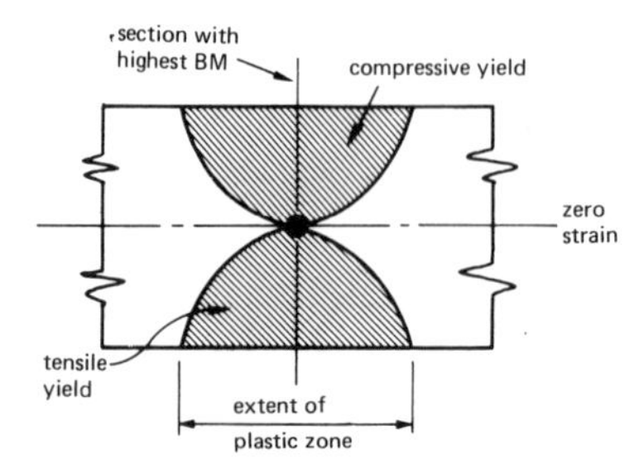

Figure 2.16

The zones cause a gradual bend in the member, rather than the sharp kink which would result from a true plastic hinge.

Since most framed structures are constructed using low shape factor sections, the errors from the idealisations are not great. The fact that it is not practically possible to develop the full plastic moment of the section is balanced by ignoring the hidden reserve in strength due to strain hardening.

2.5 FACTORS WHICH CAN ALTER THE PLASTIC MOMENT

2.5.1 Axial Force

Columns may have to carry significant axial forces in addition to bending moments. The axial force, P, moves the axis of zero strain as in figure 2.17. To simplify the mathematics the stresses have been replaced by two equivalent distributions. The stresses in **A** are assumed wholly due to the axial force, that is

$$P = bz\sigma_y \tag{2.15}$$

the stresses in **M** are caused by the changed plastic moment $M_p{}'$

$$M_p{}' = 2\,\frac{(d-z)}{2} \times b \times \sigma_y \left(\frac{d-z}{4} + \frac{z}{2}\right)$$

$$= (d-z) \times b \times \sigma_y \left(\frac{d+z}{4}\right)$$

$$= \left(\frac{d^2 - z^2}{4}\right) b\sigma_y$$

$$= \frac{bd^2}{4}\,\sigma_y \left[1 - \frac{z}{d}\right]^2$$

from equation 2.5, $bd^2\sigma_y/4$ is the plastic moment M_p of the rectangular section, so that

$$\frac{M_p{}'}{M_p} = 1 - \left(\frac{z}{d}\right)^2 \tag{2.16}$$

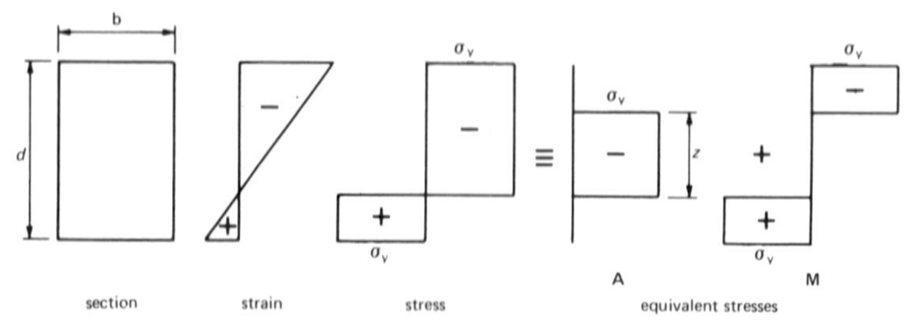

Figure 2.17

The maximum axial force P_p, which the section can carry ignoring buckling is called the *squash load* and is given by

$$P_p = bd\sigma_y$$

so that

$$\frac{P}{P_p} = \frac{z}{d}$$

and

$$\frac{M_p'}{M_p} = 1 - \left(\frac{P}{P_p}\right)^2 \tag{2.17}$$

Equation 2.17 shows that both tensile and compressive forces reduce the plastic moment because the reduction term is $(P/P_p)^2$. Equation 2.17 is plotted in figure 2.18.

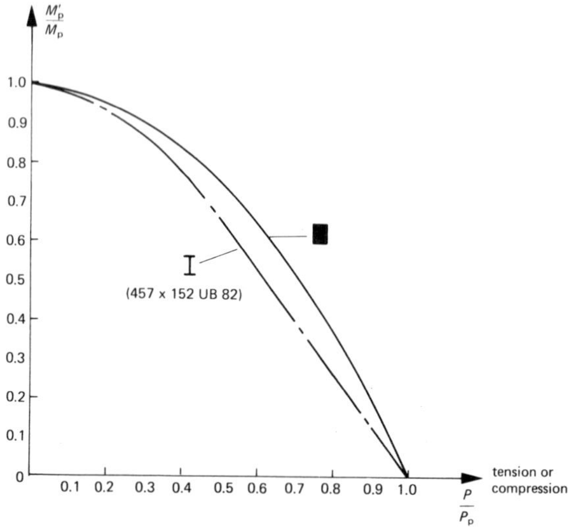

Figure 2.18

A similar result can be obtained for an I-section, but the two cases, where the zero strain axis lies in the web or the flange of the beam, must be considered separately. The equations for the more common case of bending about the y-axis are

(1) *with the zero strain axis in the web*

$$M_p' = M_p - \left(\frac{A^2}{4t_w}\right) n^2 \sigma_y \qquad (2.18)$$

where A is the total cross-sectional area and n is P/P_p. Equation 2.18 is valid while

$$\frac{P}{P_p} \leqslant 1 - \frac{2bt_f}{A} \qquad (2.19)$$

(2) *with the zero strain axis in the flange*

$$M_p' = \left[\frac{A^2}{4b}(1-n)\left(\frac{2bd}{A} - 1 + n\right)\right]\sigma_y \qquad (2.20)$$

Equations 2.18 and 2.20 are also plotted in figure 2.19 for a typical I-beam (457 x 152 UB 82). Details of the mathematics can be found in several texts. [3, 4]

In all symmetric sections the axial forces reduce the effective plastic moment irrespective of the sign of the force. In asymmetric sections the situation is more complicated, in certain circumstances it is possible to achieve an increase in plastic moment. Horne [3] has shown how this is possible.

2.5.2 Shear Force

Except in regions of constant bending moment, any section must carry a bending moment and a shear force, N. This means there is usually a combination of direct stresses σ (from the moment) and shear stresses τ. In these circumstances it is necessary to use a yield criterion to determine the start of yield. The Tresca and Von Mises criteria are the most common for ductile materials, and are described in appendix A. Where there is a combination of shear and direct stresses, both criteria require that

$$\left(\frac{\sigma}{\sigma_y}\right)^2 + \left(\frac{\tau}{\tau_y}\right)^2 = 1 \qquad (2.21)$$

for yield to occur, where τ_y is the yield stress in pure shear. Consequently the individual stresses, σ and τ, cannot reach their full yield value unless the other is zero. Thus when yield occurs at a section in bending, a situation as in figure 2.19 arises. (The section shown represents a general structural section, since they usually have a rectangular web.)

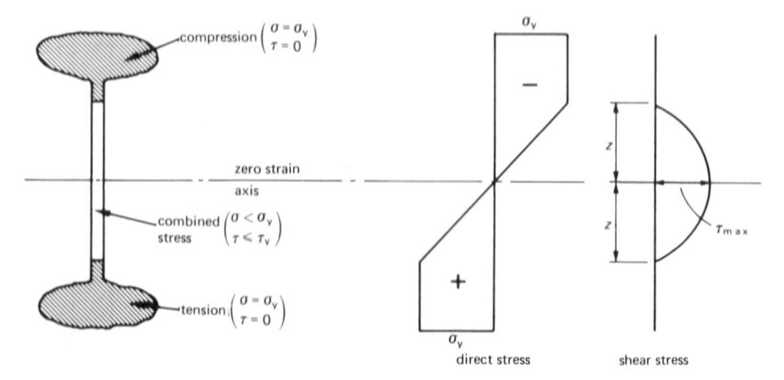

Figure 2.19

It is obvious that the direct stress distribution required for the full plastic moment can never develop. It is reasonable to assume that the shear stress distribution is parabolic, with $\tau_{max} = \tau_y$ when the plastic hinge forms (elastic analysis of a rectangle gives a parabolic distribution of shear stress). For a rectangular section in this situation, the reduced plastic moment is

$$M_p' = b\left(\frac{d^2}{4} - \frac{z^2}{3}\right)\sigma_y$$

from the stress distribution in figure 2.20. Also

$$N = \frac{4}{3} bz\tau_y \tag{2.22}$$

Since $M_p = bd^2/4\sigma_y$ and $N_p = bd\tau_y$ (yield shear stress over whole section),

$$\frac{M_p'}{M_p} = 1 - \frac{3}{4}\left(\frac{N}{N_p}\right)^2 \tag{2.23}$$

Notice that z_{max} is $d/2$. If that is substituted into equation 2.22 it can be seen that equation 2.23 only holds if

$$\frac{N}{N_p} < \frac{2}{3} \tag{2.24}$$

Equation 2.23 is plotted in figure 2.20. I-sections can be treated in a similar manner. The reduced moment is given by

$$M_p' = M_p - \frac{3}{4}\left(\frac{N}{N_{pw}}\right)^2 M_{pw} \tag{2.25}$$

if $N/N_{pw} < 2/3$, where

$$N_{pw} = t_w(d - 2t_f)\tau_y$$

the maximum shear force in the web, and

$$M_{pw} = \frac{t_w(d - 2t_f)^2}{4} \sigma_y$$

the plastic moment of the web. This is also plotted in figure 2.20.

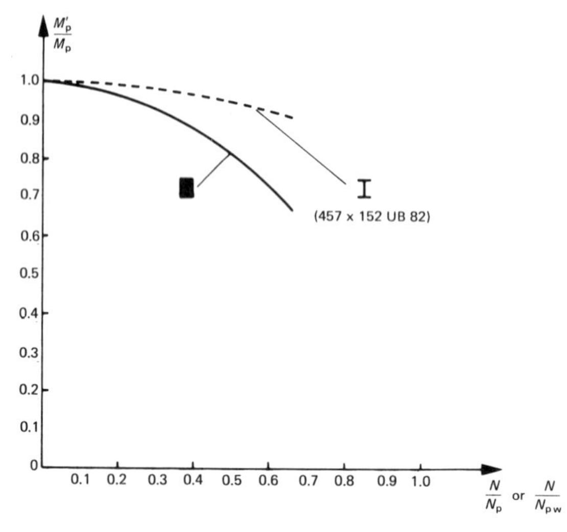

Figure 2.20

2.5.3 How Important are these Factors?

Usually they are not important. In low-rise structures axial forces are too small to have any significant effect on the plastic moments. In high-rise structures they are more important, although the problem is more likely to be instability (see chapter 6), and allowance must be made for them. As can be seen from figure 2.20 shear forces cause smaller reductions in plastic moment than axial forces, and need only be considered in the rare cases when they are exceptionally large.

2.6 SUMMARY

This chapter has been mainly concerned with yielding at a particular section of a member in bending. When full yielding has taken place and unrestrained plastic flow occurs, a *plastic hinge* has formed. The maximum bending moment that the section can withstand is the *plastic moment.* The plastic moment is a geometric property of the section. The plastic hinge and moment concepts are idealisations of the true section behaviour.

Axial and shear forces can affect the magnitude of the effective plastic moment, although they are not usually significant factors.

The formation of a plastic hinge in a framed structure is the equivalent of a member yielding in a truss type structure. The fact that part of a section in bending is in compression could cause local buckling problems but standard structural sections are proportioned to prevent this happening. The plastic moments and axial load effects on the plastic moments of the standard sections are tabulated in various handbooks [2], but for built-up sections the methods outlined in this chapter must be used.

2.7 PROBLEMS

2.1 Calculate the plastic moment of a pair of identical plates (width b, thickness t). The plates are parallel with a clear distance D between them ($D \gg t$). The axis of bending is parallel to the plates.

Plates (150 mm x 12 mm) are welded to the flanges of an I-beam (S = = 1800 cm^3, overall depth = 467.4 mm). Find the plastic moment of the combined section, assuming a yield stress of 250 N/mm^2.

2.2 Show that the plastic moduli for bending about the y-and z-axes of the channel section in figure 2.21 are

$$S = 5D^2t \qquad \text{for bending about the } y\text{-axis}$$

$$S = 1.75D^2t \qquad \text{for bending about the } z\text{-axis}$$

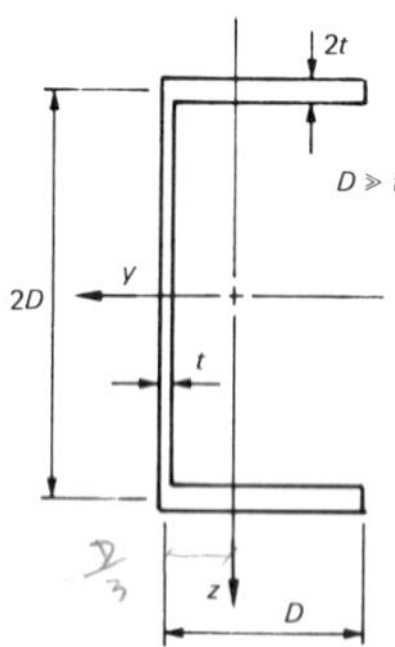

Figure 2.21

2.3 Find the plastic moment of the section shown in figure 2.22. Assume the yield stress is 350 N/mm^2.

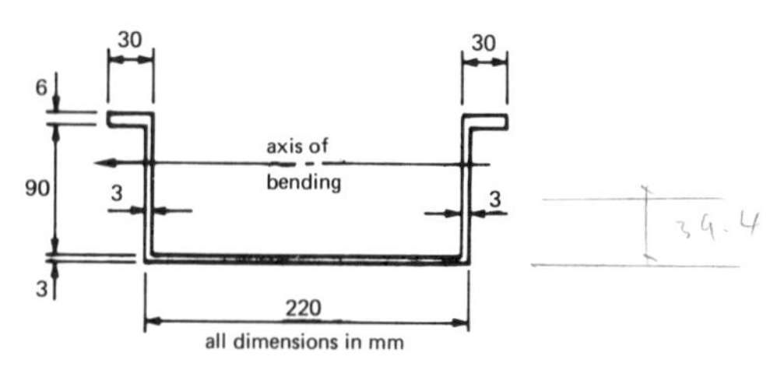

Figure 2.22

2.4 Find the plastic moments of

√(a) a solid circular section, diameter D, bending about a diameter.

(b) a thin walled square section, side length d, thickness t ($d \gg t$), bending about an axis parallel to two sides.

√(c) the same section as in (b) bending about a diagonal.

(d) a thin walled equilateral triangle section, side length a, thickness t ($a \gg t$), bending about an axis parallel to one side.

2.5 Derive the expressions in equations 2.18 and 2.20 for the reduced plastic moment, due to an axial force, of the I-beam in figure 2.11. Derive also the corresponding expressions for bending about the z-axis.

The beam 533 x 21 UB 122 has the properties

d = 544.6 mm $\qquad t_w$ = 12.8 mm

b = 211.9 mm $\qquad t_f$ = 21.3 mm

A = 155.8 cm^2 $\qquad \sigma_y$ = 250 N/mm^2

S = 3202 cm^3 for bending about the y-axis

S = 500.6 cm^3 for bending about the z-axis

Find the reduced plastic moments when it carries axial forces of 1150 kN and 2300 kN.

2.6 Derive the expression 2.25 for the reduced plastic moment, due to a shear force, of an I-beam.

Show that if the shear stress τ is assumed to be constant over the web ($\tau \leqslant \tau_y$) so that the shear force $N = t_w(d - 2t_f)\tau$, the reduced plastic moment is

$$M_p' = M_p - \left\{1 - \sqrt{\left[1 - \left(\frac{N}{N_{pw}}\right)^2\right]}\right\} M_{pw}$$

where the various terms have the same meaning as in section 2.5.2.

3 COLLAPSE OF SIMPLE FRAMES

3.1 INTRODUCTION

It was shown in the previous chapter that the formation of a plastic hinge in a framed structure is equivalent to a member yielding in a truss. The purpose of the first part of this chapter is to show what happens in a frame as the load is increased until failure occurs. This means going through an example in the same way as in chapter 1. This example will also be used to illustrate some important theorems which are essential to plastic analysis.

The second part of the chapter introduces two powerful methods for determining the collapse loads of beams and portal frames.

3.2 THE BEHAVIOUR OF A PORTAL FRAME UNDER INCREASING LOAD

A portal frame is shown in figure 3.1, carrying loads λV and λH. The relative magnitudes are determined by V and H, the absolute magnitudes by the load factor λ. Assume initially that $V = H = 1.0$. The behaviour of the frame, as λ is increased, is summarised in figure 3.2.

Initially the frame is elastic everywhere, and an elastic analysis gives a bending moment diagram (BMD) as shown in stage 1. (The slope deflection

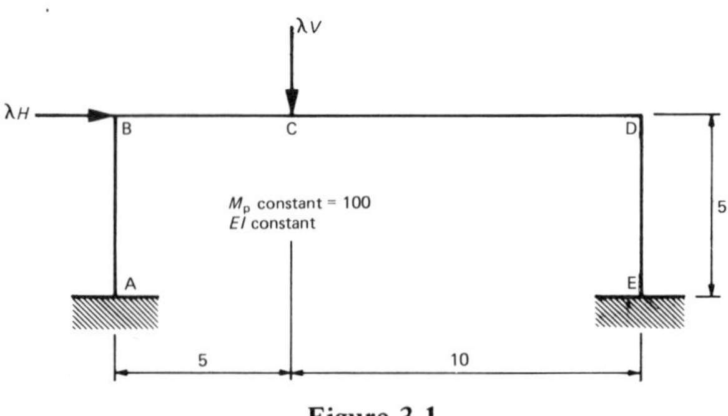

Figure 3.1

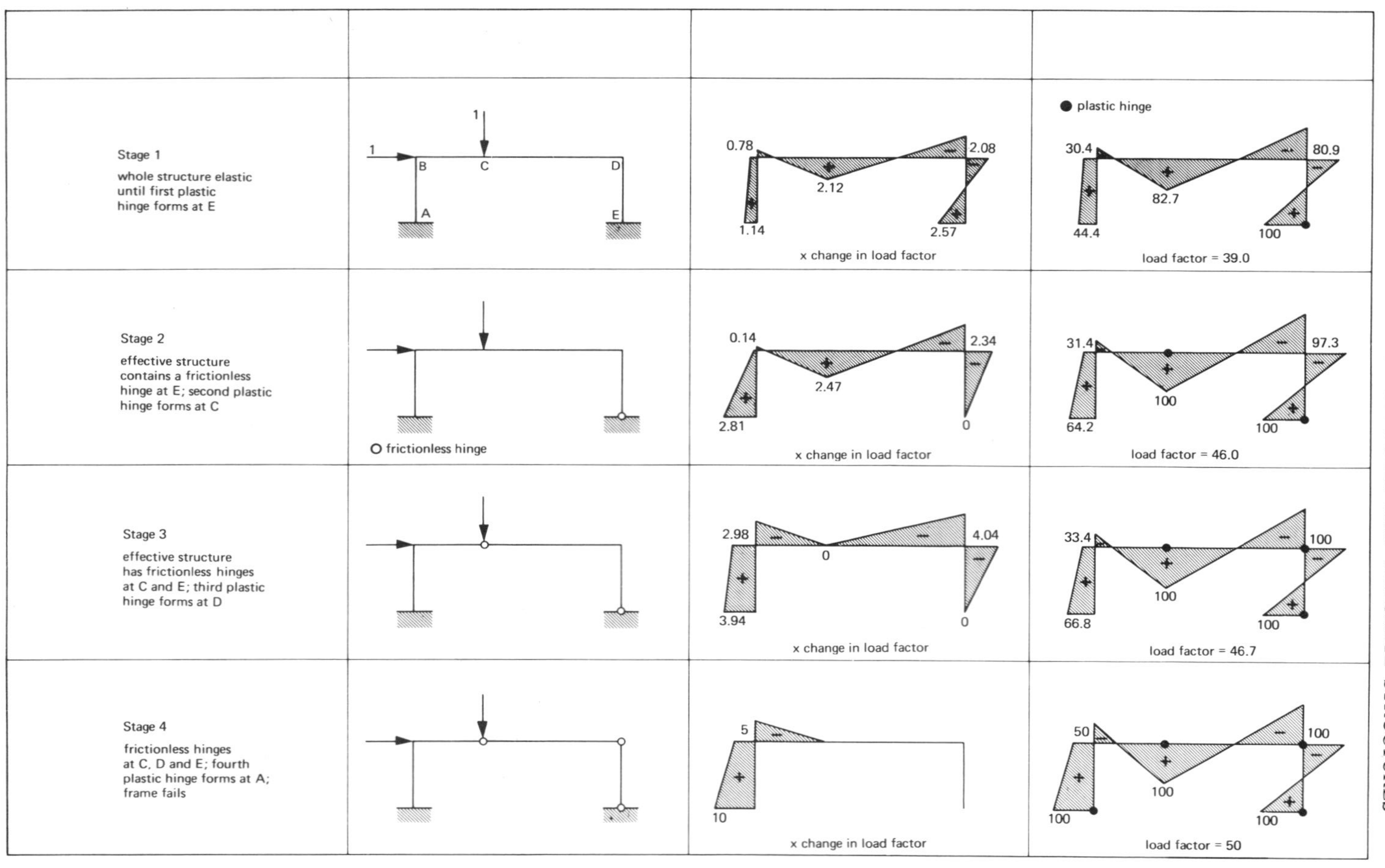

Stage 1
whole structure elastic until first plastic hinge forms at E
1
1
A
B
C
D
E
0.78
2.12
2.08
1.14
2.57
x change in load factor
● plastic hinge
30.4
82.7
80.9
44.4
100
load factor = 39.0
Stage 2
effective structure contains a frictionless hinge at E; second plastic hinge forms at C
O frictionless hinge
0.14
2.47
2.34
2.81
0
x change in load factor
31.4
97.3
100
64.2
100
load factor = 46.0
Stage 3
effective structure has frictionless hinges at C and E; third plastic hinge forms at D
2.98
0
4.04
3.94
0
x change in load factor
33.4
100
100
66.8
100
load factor = 46.7
Stage 4
frictionless hinges at C, D and E; fourth plastic hinge forms at A; frame fails
5
10
x change in load factor
50
100
100
100
100
load factor = 50

Figure 3.2

method and a small computer were in fact used.) When $\lambda = 39.0$ the largest bending moment (BM), at the foot of the right-hand column (point E) becomes equal to the plastic moment and a plastic hinge forms. Of course, the whole structure apart from E is still elastic and remains so when λ is increased above 39.0. When λ is increased, E behaves like a hinge in that it can rotate freely, but the BM must remain equal to the plastic moment.

Stage 2 shows the effective structure which resists the loads when λ is greater than 39.0, it is simply the original frame with a frictionless hinge at E. This structure can be analysed by the same elastic method as in stage 1. The result of the analysis is the *change* in BMs. To get the total moments it is necessary to add the change in BMs to the BMs when $\lambda = 39.0$. (Notice that the frictionless hinge at E ensures that the change in BM at E is zero so that the total moment remains equal to the plastic moment.) The maximum moment is under the vertical load, point C

$$M_C = 82.7 + 2.47\lambda'$$

where λ' = change in λ and this equals the plastic moment (100) when $\lambda' = 7.0$, and $\lambda = 46.0$.

As stage 3 indicates, from now on there are two hinges in the effective structure, but it can still be analysed elastically. Very rapidly, however, a further hinge forms at D when $\lambda = 46.7$.

It is worth noting two points about the total BMs shown in the right-hand column of figure 3.2.

(1) *Equilibrium condition* – the distribution of BMs is in equilibrium with the applied loads. (This is the basis of the slope deflection analysis.)

(2) *Yield condition* – the BMs nowhere exceed the plastic moment of the members. (Inspection of the total BMs at each stage shows this.)

The process can be continued as in stage 4 with three frictionless hinges, until at $\lambda = 50$ a fourth plastic hinge forms. Any attempt to continue the process using an effective structure with four frictionless hinges is impossible, the equations become singular and cannot be solved. In fact, the structure becomes a mechanism and is on the point of collapse when the fourth hinge forms. The BMD at this point, see stage 4, satisfies the equilibrium and yield conditions, and the structure also satisfies:

(3) *Mechanism condition* – there are sufficient plastic hinges for the structure to become a mechanism.

This load factor is called the ***collapse load factor***, λ_c.

The slope deflection method was most convenient for this analysis because it also gave the deflections of the structure. The horizontal deflection at the top of the columns and the vertical deflections under the vertical load are plotted in figure 3.3 against the load factor λ. The shape of the graphs is very similar to the load deflection graph of the truss in figure 1.13, showing a sudden

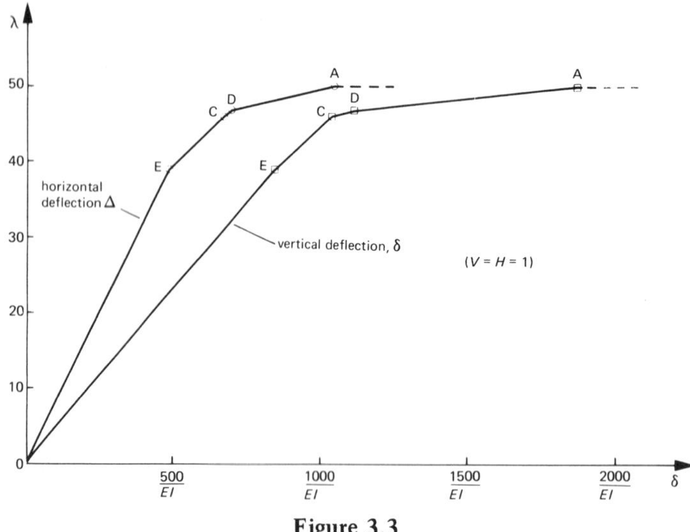

Figure 3.3

reduction in slope (stiffness) as each hinge forms, until at collapse the slope is zero. There are other similarities to the truss. The number of hinges at collapse is one more than the degree of redundancy of the frame. (Four hinges at collapse, and the test in appendix B shows that the degree of redundancy is 3.) The formation of a new hinge causes a drop in the stiffness of the frame and the removal of one degree of redundancy. Comparison of the BM values at the end of each stage reveals considerable redistribution.

As collapse develops, the bending moment distribution remains the same as at the point of collapse, while the deflections remain 'small'. (Obviously if the moment at a plastic hinge were to get smaller the hinge and mechanism would cease to exist!) The way collapse develops is shown in figure 3.4. When the hinge rotations are small, the shape of the structure is only slightly altered, but with large rotations the shape is completely different.

This theoretical behaviour is confirmed by experiments. Figure 3.5 shows three model portal frames which have been tested with various combinations of

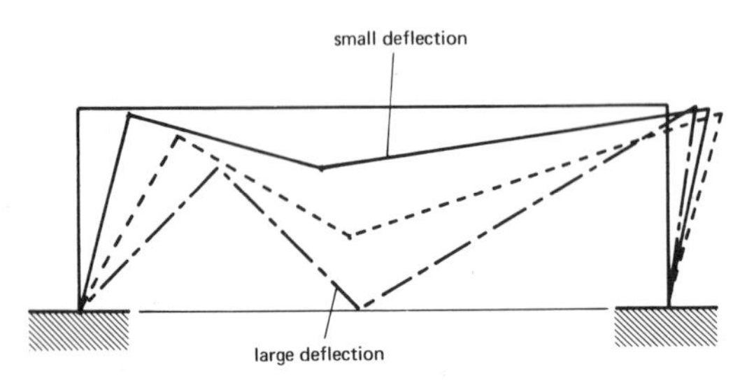

Figure 3.4

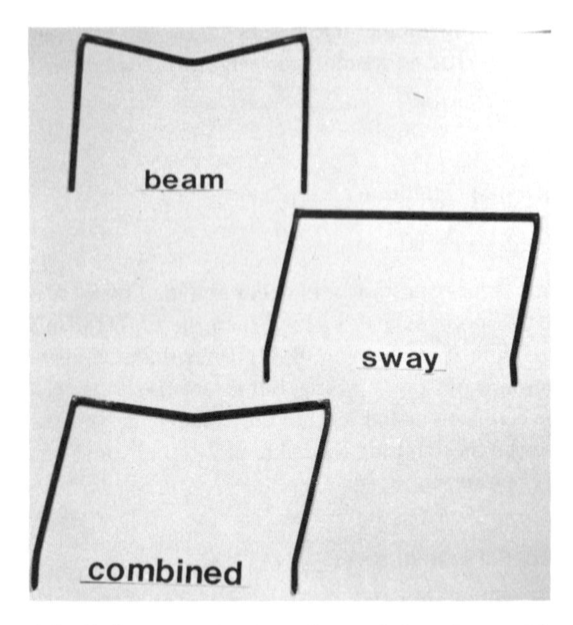

Figure 3.5 Collapse mechanisms for model steel portal frames

loading. The loadings have caused different collapse mechanisms but the plastic hinges can clearly be seen in each case.

The mechanism in stage 4 of figure 3.2 is the actual collapse mechanism of the structure. It should be possible to guess some other mechanism and work backwards to find the value of λ at which it would occur. Consider the mechanism in figure 3.6a. It has the same number of hinges, but one is in a

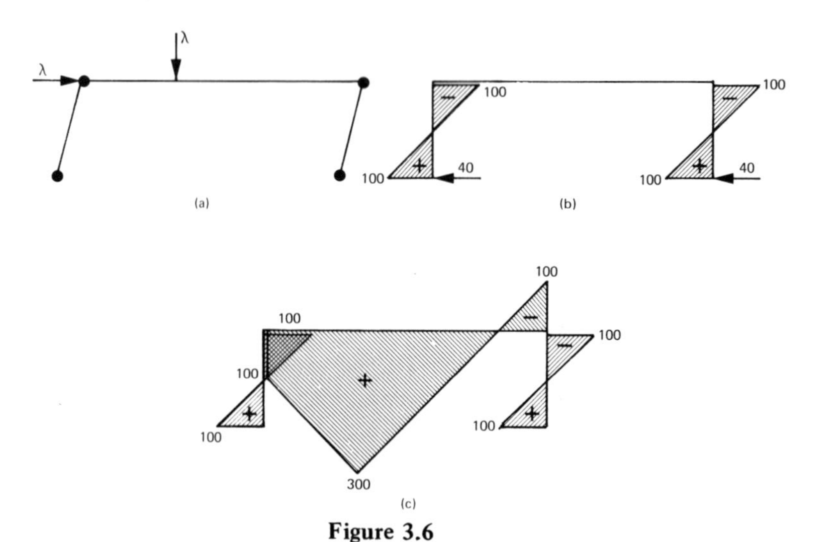

Figure 3.6

different position. Using the method in appendix C the BMD in the columns is as shown in figure 3.6b. The horizontal reaction at the foot of both columns is

$$H = \frac{100 + 100}{5} = 40$$

So that for horizontal equilibrium

$$\lambda = 40 + 40 = 80$$

This means that the mechanism would occur at a load factor $\lambda = 80$, which is higher than the true collapse load factor. When the BMD for this mechanism is completed as in figure 3.6c, there are BMs greater than the plastic moment. This guessed mechanism is obviously wrong, but interestingly its corresponding BMD still satisfies the equilibrium and mechanism conditions. This turns out to be true for any guessed mechanism. The value of λ found for the guessed mechanism is called an *upper bound* to λ_c.

3.3 THE THEOREMS OF PLASTIC ANALYSIS

The information about the three conditions from the previous section is summarised below, where the arrows indicate conditions which have been satisfied.

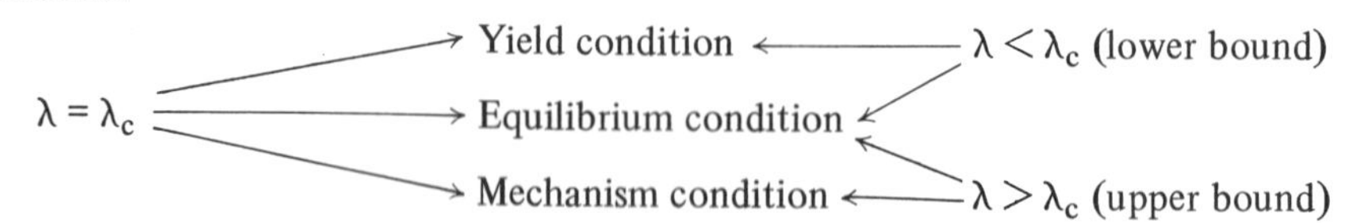

It can be proved [4] that the three situations are always true, and they have been embodied in the three essential theorems of plastic analysis.

3.3.1 Lower Bound Theorem

If, in a structure subjected to loading defined by a positive load factor λ, a BM distribution satisfying the equilibrium and yield conditions can be found, then λ is less than or equal to the collapse load factor λ_c.

3.3.2 Upper Bound Theorem

If, in a structure subjected to loading defined by a positive load factor λ, a BM distribution satisfying the equilibrium and mechanism conditions can be found, then λ is greater than or equal to the collapse load factor λ_c.

3.3.3 Uniqueness Theorem

If a structure is subjected to loading, defined by a positive load factor λ, such that the resulting BM distribution satisfies the three collapse conditions, then λ equals λ_c. It is impossible to obtain a BM distribution, at any other load factor, which satisfies all three conditions simultaneously.

The object of plastic analysis is to find collapse loads directly. As was hinted in section 1.3.2.4 the calculations are easier than those for elastic analysis, for example there are no simultaneous equations to solve. The rest of this chapter and the whole of the next are devoted to methods of finding the collapse loads of framed structures. These methods are based on the theorems presented in this section.

3.4 THE NUMBER OF HINGES REQUIRED IN A MECHANISM

As shown in section 3.2 and also at length in chapter 1, generally the number of hinges, n, in a collapse mechanism is one more than the degree of redundancy, r

$$n = r + 1 \tag{3.1}$$

This is a useful guide to make sure that a mechanism has been created. Unfortunately, there are two exceptions to this rule which must be identified.

The frame in figure 3.1 has been re-analysed for the loadings $H = 0.5$, $V = 3$ and $H = 4$, $V = 1$. The results are presented in figures 3.7 and 3.8 respectively. In the first case $\lambda_c = 20.0$, but there are only three plastic hinges, one less than required by equation 3.1, at collapse. This situation is called *partial collapse* because that is exactly what happens in the structure. In the example the beam has failed prematurely. In the second case the last two plastic hinges form simultaneously so that there are five hinges at collapse. This is called *over collapse*.

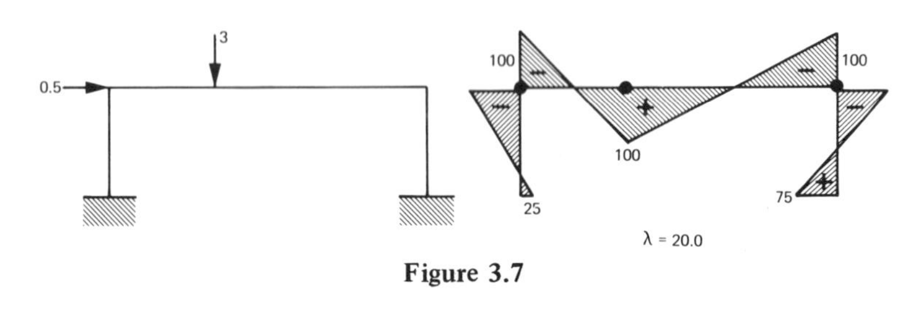

Figure 3.7

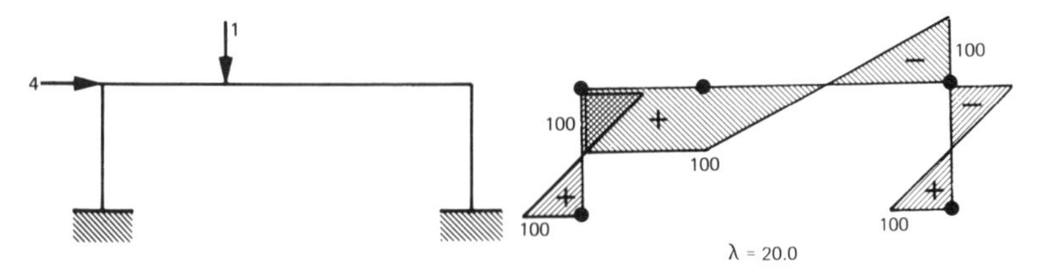

Figure 3.8

3.5 FREE AND REACTANT BM METHOD FOR FINDING COLLAPSE LOADS

This method is limited to beams, where the BMs at collapse can be written down using some simple rules. The method will be illustrated by a series of examples. Each example will bring out a further aspect of the method. It is therefore recommended that the reader works through the examples in order.

3.5.1 Simply Supported Beam

A simply supported beam with a central point load is shown in figure 3.9a. It is determinate so only one plastic hinge is required for collapse. This hinge forms at the point of maximum BM to form the mechanism in figure 3.9b. Hence collapse occurs when the maximum BM equals the plastic moment of the beam

$$\frac{W_c L}{4} = M_p$$

$$W_c = \frac{4M_p}{L} \tag{3.2}$$

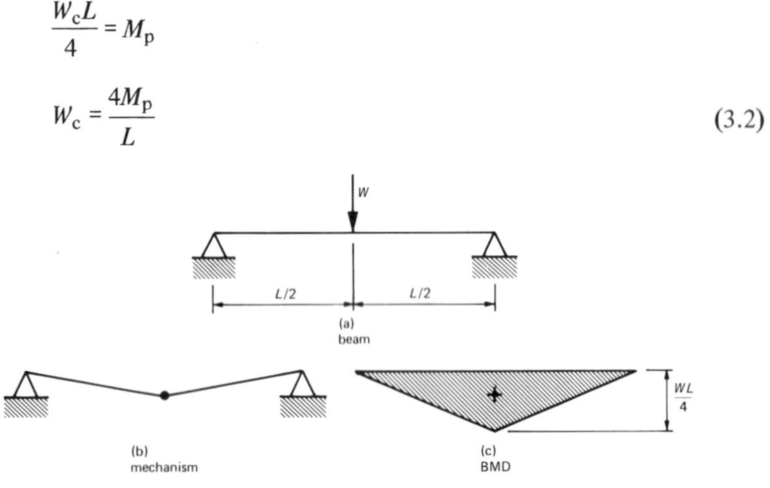

Figure 3.9

3.5.2 Beam with Fixed Ends

A beam with fixed ends and carrying a point load is shown in figure 3.10a. Figure 3.10b shows the BMD for elastic behaviour. The largest moment

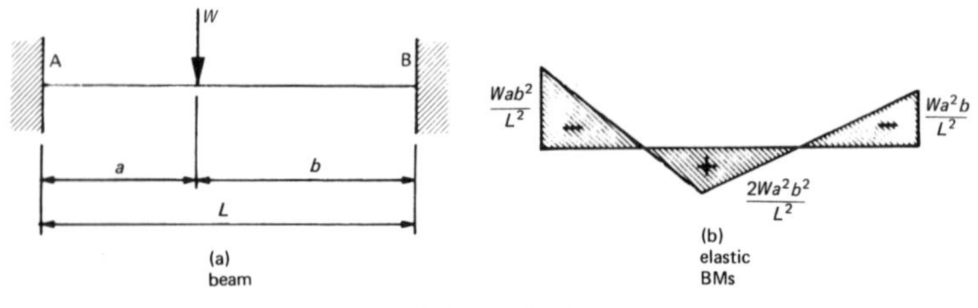

Figure 3.10

Wab^2/L^2 occurs at A. It would be possible to follow through the behaviour (as in section 3.2) by inserting a hinge first at A, and then at other points when the BM becomes equal to the plastic moment. This process would give a BMD at collapse and a collapse mechanism as in figure 3.11, when $W_c = 2M_pL/ab$.

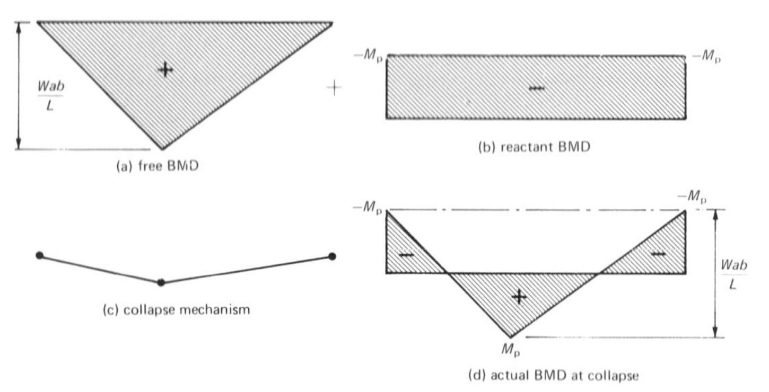

Figure 3.11

It is possible to draw the collapse mechanism and BMD without any analysis by using the simple rules

Rule 1 **In general hinges form at restrained ends of beams**
Rule 2 **Hinges form under point loads.**

Even at the point of collapse the BMD is made up from the reactant and free BMDs (see appendix C). These are shown in figures 3.11a and b for the fixed beam. The geometry of the BMD at collapse (figure 3.11d) gives the collapse load

M_p	+	M_p	$= \dfrac{Wab}{L}$
actual BM under point load		reactant BM at point load	

that is

$$W_c = \frac{2M_pL}{ab} \tag{3.3}$$

This can be extended to other problems with only some complication in the BMD geometry, as illustrated in the next example.

3.5.3 Propped Cantilever

A propped cantilever with point load is shown in figure 3.12a. Using rules 1 and 2 of section 3.5.2 the collapse BMD (figure 3.12b) can be drawn. Since this is

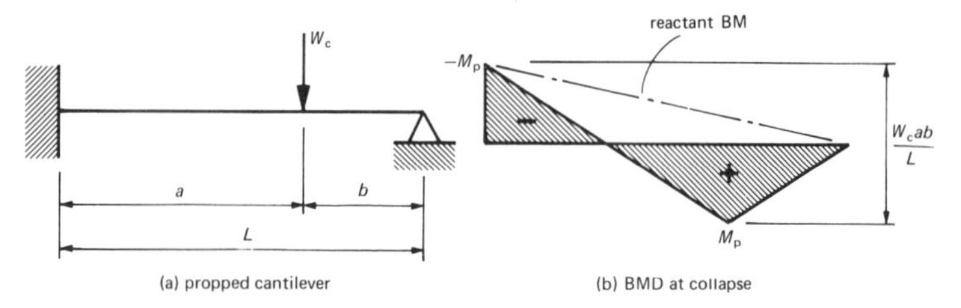

Figure 3.12

made up of the free and reactant BMDs, the geometry gives

M_p	+	$\dfrac{bM_p}{L}$	=	$\dfrac{W_c ab}{L}$
actual BM under load		reactant BM under load		free BM under load

$$W_c = \frac{M_p(L+b)}{ab} \tag{3.4}$$

The only difference between this and the previous example is the need to use similar triangles to find the reactant BM under the load.

3.5.4 Continuous Beam

Continuous beams can be analysed in a similar way. There are two further points to consider. Firstly, each span may have a different section and thus plastic moment. At a support between spans the BM is common to the beam on both sides of the support. Thus when a plastic hinge forms at a support the value of the BM at that support is decided by

Rule 3 **At a support the plastic hinge forms at the plastic moment of the weaker member.**

Secondly, it is unlikely that every span will fail simultaneously. Each span must be checked individually. The span, or spans, with the lowest collapse load determines the collapse of the whole beam. This is a good example of partial collapse.

A continuous beam and its BMD at collapse are shown in figure 3.13. The plastic hinges under the point loads are not shown, since not all of them will have formed when collapse occurs. Each span must be checked in turn for collapse.

If spans AB and CD collapse first the BMD at collapse will be as in figure 3.14. The problem is identical to the propped cantilever in section 3.5.3.

$$W_c = \frac{600(8+4)}{4 \times 4} = 450 \text{ kN}$$

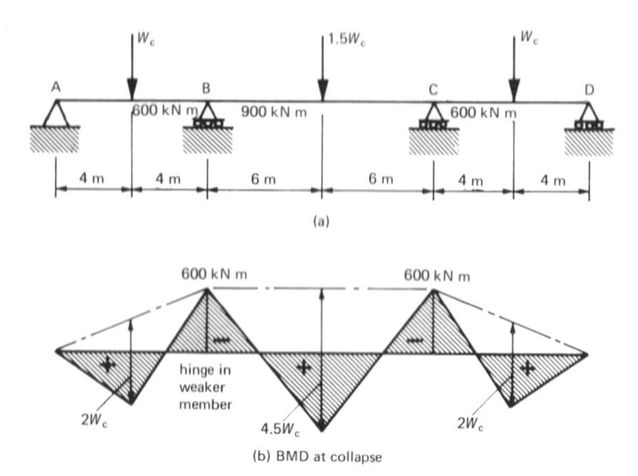

(b) BMD at collapse

Figure 3.13

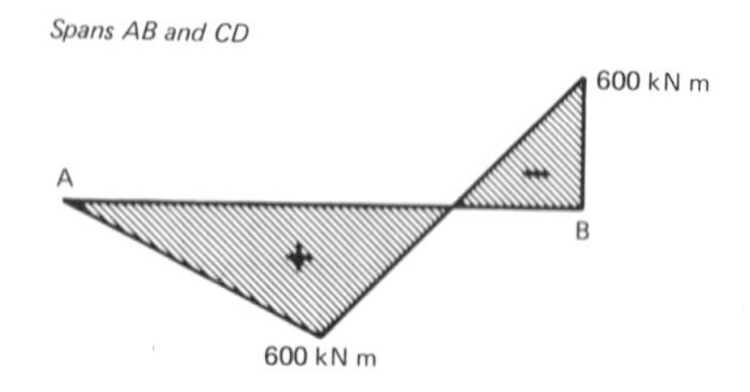

Figure 3.14

The collapse BMD for span BC is shown in figure 3.15. This is a very similar problem to the fixed end beam. From the geometry of the BMD

$$4.5W_c = 600 + 900$$

$$W_c = 333 \text{ kN}$$

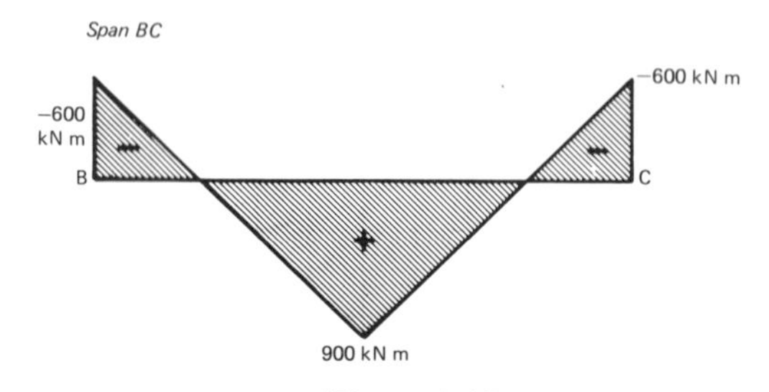

Figure 3.15

Obviously a mechanism forms first in span BC and collapse occurs when $W_c = 333$ kN. The collapse BMD for the whole beam can now be completed. It is shown in figure 3.16a. Notice that the moments under the point loads in AB and CD are smaller than the plastic moment, because a mechanism only forms in BC (figure 3.16b).

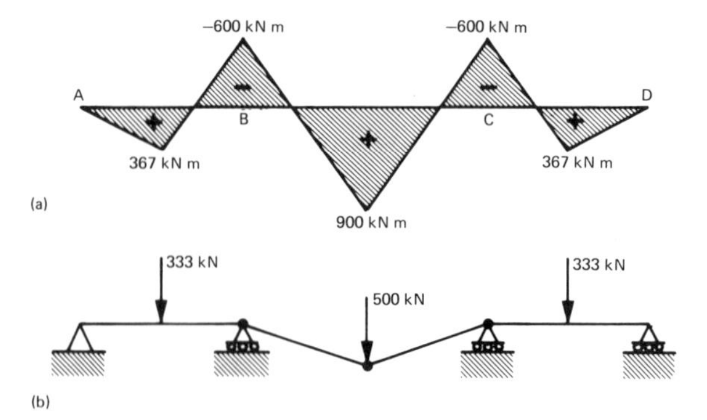

Figure 3.16

3.5.5 Spans with Distributed Loads

In all the previous sections of this chapter the beams have carried concentrated loads. In practice loads have a habit of being distributed along the length of the span. This makes things more difficult because the position of the plastic hinge within the span can no longer be decided by applying a simple rule.

A typical span and collapse BMD are shown in figure 3.17a and b. The position of the span hinge is not obvious. It occurs where the resultant BM is a maximum. The condition for this maximum is

$$\frac{dM}{dx} = 0 \tag{3.5}$$

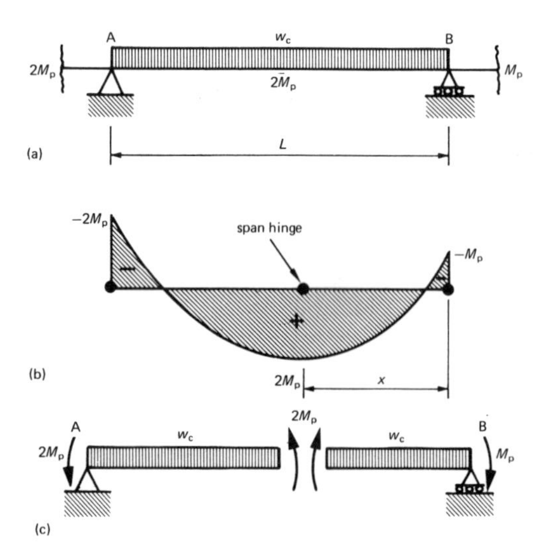

Figure 3.17

Using the relationship between BM and shear force

$$\frac{dM}{dx} = N \tag{3.6}$$

if the beam is cut at the span hinge point the free body diagrams in figure 3.17c describe the equilibrium of the beam. The magnitudes of all moments are known and the shear force must be zero at the cut (from equations 3.5 and 3.6), thus the only unknowns are w_c and x. Equilibrium of the two rigid bodies will give enough equations to find w_c and x. Taking moments about B for the right-hand section

$$-\frac{w_c x^2}{2} + 2M_p + M_p = 0$$

$$\frac{w_c x^2}{2} = 3M_p \tag{3.7}$$

Similarly for the left-hand section, taking moments about A

$$\frac{w_c(L-x)^2}{2} - 2M_p - 2M_p = 0$$

$$\frac{w_c(L-x)^2}{2} = 4M_p \tag{3.8}$$

substitution of equation 3.7 in equation 3.8 gives

$$\frac{w_c(L-x)^2}{2} = \frac{4}{3}\left(\frac{w_c x^2}{2}\right)$$

w_c can be cancelled out and the equation rearranged to give

$$3(L-x)^2 = 4x^2$$

$$3(L^2 - 2xL + x^2) = 4x^2$$

$$x^2 + 6xL - 3L^2 = 0$$

this quadratic equation can be solved to give

$$x = (-3 \pm 2\sqrt{3})L$$

The positive root gives the position of the span hinge.

$$x = (-3 + 2\sqrt{3})L = 0.464L$$

this can be substituted back into equation 3.7 to give

$$w_c = \frac{27.86M_p}{L^2}$$

This example can be generalised to any value of plastic moment at the ends of the member. In a propped cantilever the reactant moment at one end would be zero. Of course, the example above should not be followed blindly. When the reactant moments at each end of the span are equal, the maximum span moment, by symmetry, will be at midspan and the approach given for point loads can be used.

3.6 THE VIRTUAL WORK METHOD FOR FINDING COLLAPSE LOADS

The analysis of a frame must cover more possibilities than the analysis of a beam. The frame may collapse by being pushed sideways (sway) by horizontal forces, an individual beam may fail due to vertical loads, or there may be some combination of both. The free and reactant BM method relies on being able to decide the mechanism and BMD at collapse, which is more difficult in a frame.

The method of determining collapse loads based on the principle of virtual work has proved to be a powerful tool because it is readily applied to frames. The method is based on two premises.

(1) When a framed structure collapses all deformation of the structure occurs by rotation at the plastic hinges (this is what happens in tests).

(2) The principle of virtual work can be applied to these deformations.

Figure 3.18a shows the frame used in section 3.2. at the point of collapse, when the last hinge has just formed but no rotation has occurred. Apply an infinitely small horizontal displacement to the top of the left hand column; assuming no axial shortening of any members, the structure will deflect as in figure 3.18b, although the infinitely small deflections and rotations are greatly exaggerated.

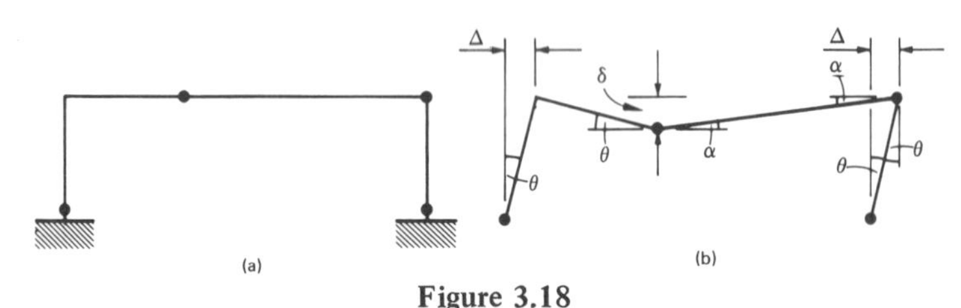

Figure 3.18

The deflections and rotations can be thought of as virtual displacements. So that

external (virtual) work, done by the applied loads $= H\Delta + V\delta$

$= \Sigma W\delta$ over all loads

internal (virtual) work, done by the applied loads $= M_p\theta + 2M_p\,(\theta + \alpha) + M_p\theta$

$= \Sigma M_p\theta$ over all plastic hinges

since the only distortions of the structure are the virtual rotations at the hinge points. At the point of collapse the structure is just in equilibrium, so that from the principle of virtual work

external (virtual) work = internal (virtual) work

that is

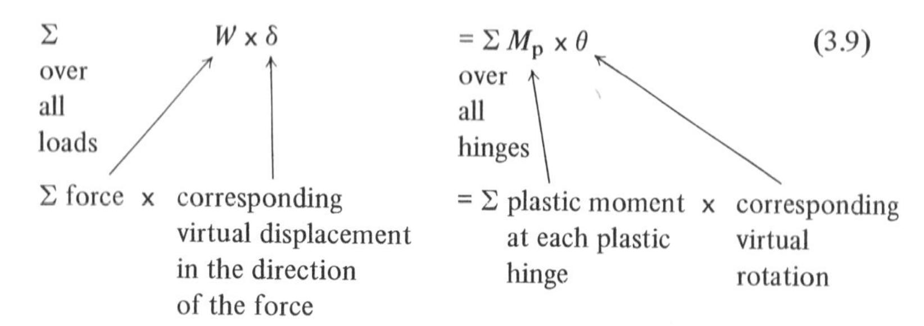

There is usually a simple relationship between the δs and the θs from the geometry of the virtual displacements, so that the loading W can be found in terms of the plastic moments M_p.

Equation 3.9 is called the *work equation.* In addition to the premises (1) and (2) above, there are two other assumptions which are implicit in the work equation.

During collapse all deformation occurs at the plastic hinges. In order to maintain the mechanism the BM cannot drop below the plastic moment at each plastic hinge, nor is it physically possible for it to increase above that value. In addition the loads remain constant as collapse occurs. Hence the additional assumptions are

(3) at collapse the BMs remain constant as the structure deforms.
(4) all axial load effects are ignored.

Therefore, as well as ignoring axial shortening, premature failure due to buckling is also ignored. This point will be considered in more detail in chapter 6.

This section has only given the bare bones of the virtual work method. Its practical application is best illustrated by some worked examples. There are four examples given, each one introducing a further aspect of the method.

3.6.1 Fixed End Beam (again)

It is worth repeating this example because it illustrates clearly the solution procedure using the virtual work method.

Problem

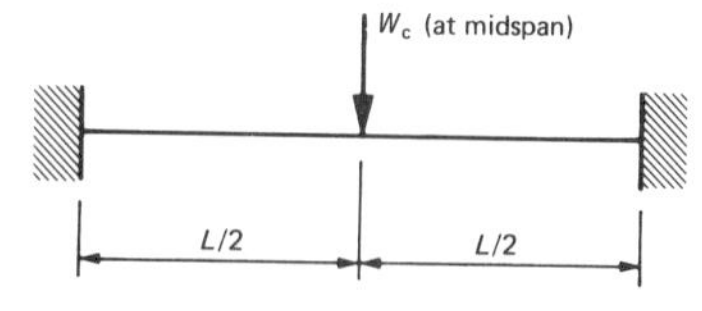

Stage 1

Collapse mechanism

$-M_p$ (hogging) $-M_p$ (hogging)

$+M_p$ (sagging)

Plastic hinges at both ends and under the load.

Stage 2

Impose virtual plastic rotations at the hinges.

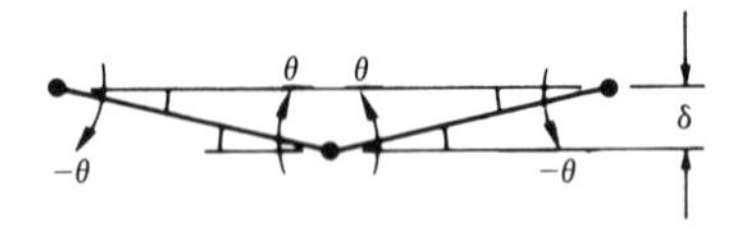

Note the signs of the rotations. If the rotation is caused by a positive (sagging) moment then it is positive.

Stage 3

Virtual work done by external load (External work) is

$$W_c \times \delta \qquad \text{(= load x virtual distance moved)}$$

Stage 4

Virtual work absorbed by hinge rotation (Internal work) is

$$-M_p \times -\theta + M_p 2\theta + -M_p \times -\theta = 4M_p\theta$$

The sign of the moments and rotations is immaterial, work absorbed is positive. Hence *no sign convention* is required.

Stage 5

Geometry of the mechanism

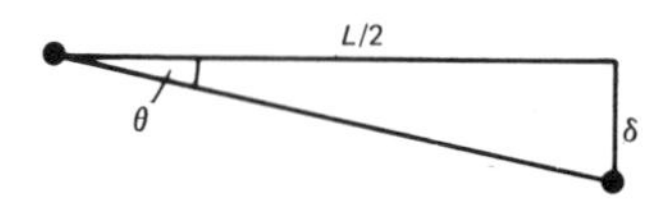

The diagram shows the left-hand half of the beam. Since all deformation occurs by plastic rotation, the beam remains straight between the hinges. $\tan\theta = \theta$ since θ is an infinitely small angle.

$$\delta = \frac{L}{2}\tan\theta = \frac{L}{2}\theta$$

Stage 6

Impose equilibrium and the geometry of the mechanism

work done = work absorbed

$$W_c\,\delta = 4M_p\theta$$

$$W_c\,\frac{L}{2}\theta = 4M_p\theta \quad \text{i.e.} \quad W_c = \frac{8M_p}{L}$$

As is to be expected the magnitude, θ, of the virtual rotation cancels from the equation. Compare this to the solution obtained in section 3.5.2, when $a = b = L/2$

$$W_c = \frac{2M_pL}{ab} = \frac{2M_pL}{(L^2/4)} = \frac{8M_p}{L}$$

It is reassuring to be able to check that the two methods give the same result for the collapse load.

There is no need to set out the calculations stage by stage. It was only done in this example to show the reasoning process required in finding the collapse load.

3.6.2 Portal Frame with Pinned Feet

As can be seen in figure 3.19 there are now horizontal and vertical forces acting on the structure. This means that *Stage 1*, deciding on the collapse mechanism, presents problems, because there are various possibilities for the mechanism. In fact a separate calculation is required for each mechanism.

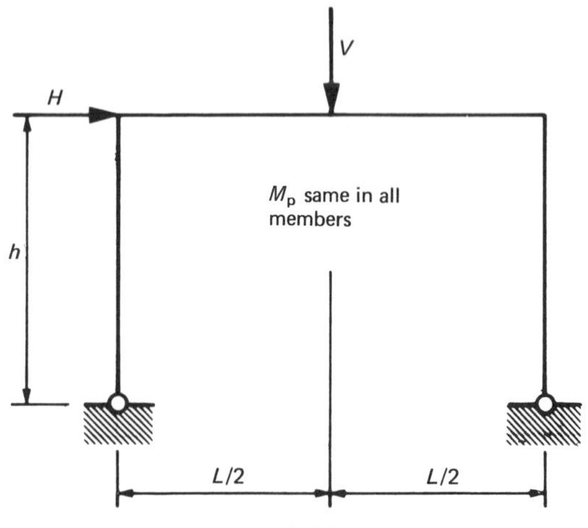

Figure 3.19

3.6.2.1 Beam Collapse

In this case collapse is caused by the action of the vertical force alone (although the horizontal force need not be zero). The plastic hinges and imposed rotations are shown in figure 3.20. The calculation is identical to the previous one.

$$V_c \frac{L}{2}\theta = 4M_p\theta \tag{3.10}$$

or

$$\frac{V_c L}{M_p} = 8 \tag{3.11}$$

Notice that the horizontal force does no work because the tops of the columns are assumed to remain stationary.

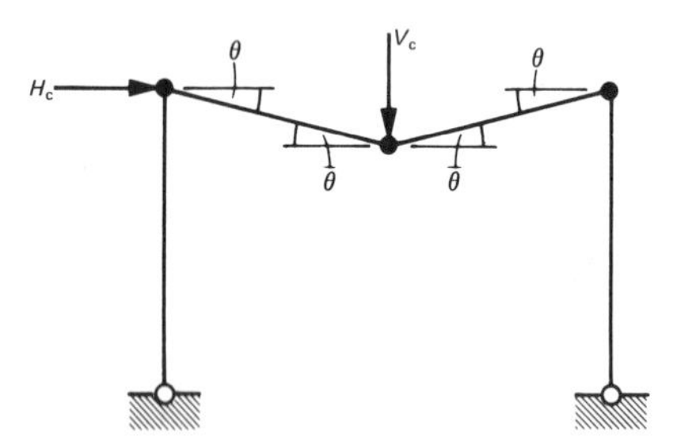

Figure 3.20

3.6.2.2 Sway Collapse

Collapse is caused by the horizontal force alone, and the structure is pushed sideways (figure 3.21). The mechanism forms when there are plastic hinges at the top of both columns. The hinges at the feet of the columns are free to rotate and are *unable* to absorb any work. (This sort of hinge is built into the frame and is assumed to be frictionless, so that it offers no resistance to rotation.)

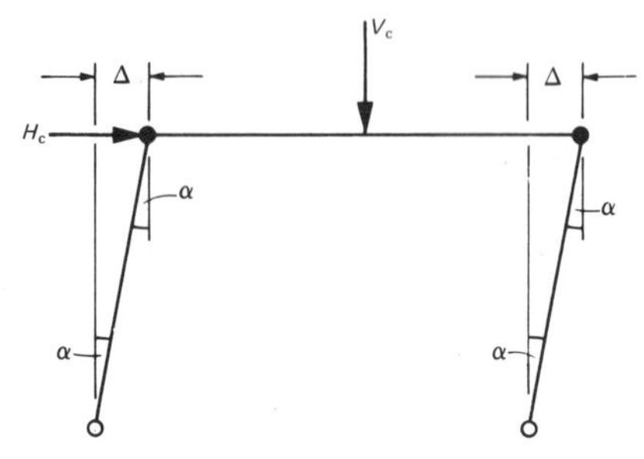

Figure 3.21

Since all deformation occurs by hinge rotation, the beam will not change length, and the top of each column moves sideways by the same (infinitely small) distance. Consequently the hinge rotation is identical in both columns.

$$\text{external work} = H_c\Delta$$
$$= H_c h\alpha$$
$$\text{internal work} = 2M_p\alpha$$

For equilibrium

$$H_c h\alpha = 2M_p\alpha \tag{3.12}$$

so that at collapse

$$\frac{H_c h}{M_p} = 2 \tag{3.13}$$

In this mechanism the vertical force does no work because the tops of the columns are assumed to move sideways but not downwards.

3.6.2.3 Combined Beam and Sway Collapse

This would seem to be a fairly obvious possibility, but does it have any significance? In fact it is only significant in the special case when the basic virtual rotation is the same in the beam and sway mechanisms, that is, $\theta = \alpha$. What happens is shown in figure 3.22

The rotation of the left-hand column and the left-hand end of the beam are such that the members remain at right-angles to each other where they connect together. This means that there has been no plastic rotation at that point, so that

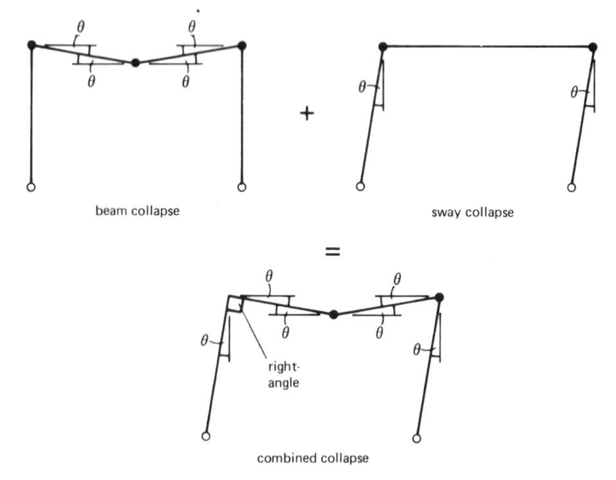

Figure 3.22

the plastic hinge (and its work-absorbing capacity) at the left-hand beam–column intersection is removed from the beam and sway mechanisms. The work equation for the combined collapse mechanism is obtained by adding the work equations (equations 3.10 and 3.12) for the beam and sway collapse and then substracting from the internal work, the work done by the hinge which disappears. Thus

$$\frac{V_c L\theta}{2} + H_c L\theta = 4M_p\theta + 2M_p\theta - \underset{\text{(beam)}}{M_p\theta} - \underset{\text{(sway)}}{M_p\theta}$$

external work, beam mechanism	+	external work, sway mechanism	=	internal work, beam mechanism	+	internal work, sway mechanism	–	internal work at hinge which disappears

that is

$$\frac{V_c L\theta}{2} + H_c h\theta = 4M_p\theta \tag{3.14}$$

$$\frac{1}{2}\frac{V_c L}{M_p} + \frac{H_c h}{M_p} = 4 \tag{3.15}$$

The concept of combining mechanisms to remove plastic hinges is very important and must be thoroughly understood. It will be used far more extensively later for analysing more complicated structures.

3.6.2.4 Which is the 'Most Likely' Mechanism?

This is a hard question to answer, because the actual collapse mechanism depends on the relative values of the forces H and V. The equations 3.11, 3.13 and 3.15 determine the loads at which the mechanisms form. They can be plotted on a graph with axes VL/M_p and Hh/M_p, as shown in figure 3.23, and

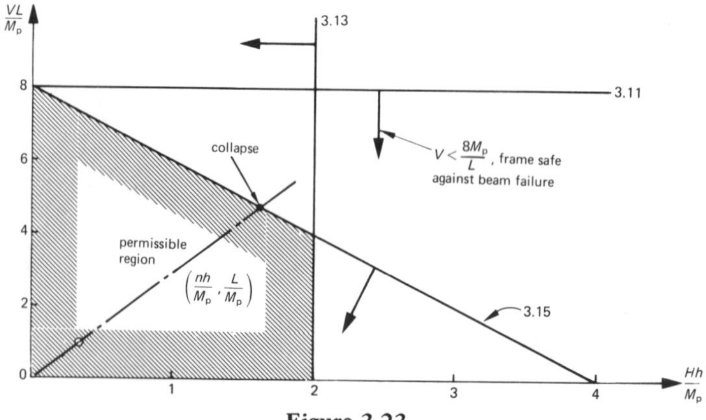

Figure 3.23

represent three straight lines. The graph is called an *interaction diagram*, (ID).

The interaction diagram gives much information about the frame. Equation 3.11 states that beam collapse will occur when $VL/M_p = 8$. If V is less than $8M_p/L$ the frame is safe against beam failure. Similar arguments can be used for the other mechanisms, and the arrows in the ID indicate safety. The shaded area, called the *permissible region*, (PR), indicates combinations of V and H which are safe against collapse by any of the possible mechanisms. If V and H are such that the corresponding point lies on the boundary of the PR then collapse will occur. In this case the boundary is made up of the sloping line (equation 3.15) representing the combined mechanism and the vertical line (equation 3.13) representing sway.

Suppose the forces are in the ratio $V/H = 1/n$, how will the frame collapse? Assume that $V = 1$, so that $H = n$. Plot on the ID the point $(nh/M_p, L/M_p)$. The point at which the line through the origin and $(nh/M_p, L/M_p)$ cuts the boundary of the PR gives the collapse mechanism (combined, as shown) and the values of V and H at collapse (as shown, $V = 4.75M_p/L, H = 1.625M_p/h$).

The axes of the ID need not be the quantities used in this example. The most convenient values should be used. An alternative is given in the next section.

3.6.3 Fixed Foot Portal Frame

Apart from the fixed feet the major differences between this and the previous example are the more complicated geometry and different plastic moments for the members. These must be accounted for in the calculations.

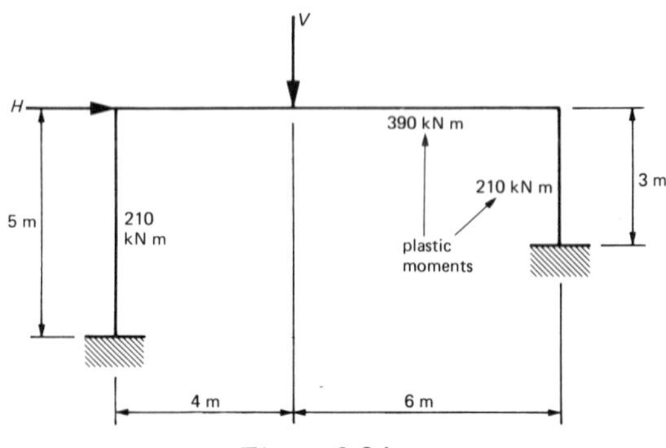

Figure 3.24

(a) Beam Mechanism

The geometry of the mechanism (see p. 56) requires

$$\delta = 4\theta = 6\alpha$$

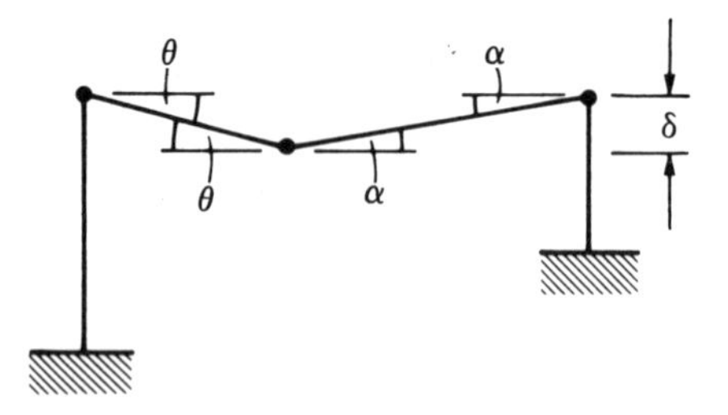

that is

$$\alpha = \frac{2}{3}\,\theta$$

At a connection between two members the plastic hinge forms at a BM equal to the plastic moment of the weaker member. The work equation is thus

$$V\delta = 210\theta + 390(\theta + \alpha) + 210\alpha$$

$$= 600\,(\theta + \alpha)$$

substituting for δ and α gives

$$V \times 4\theta = 600\left(1 + \frac{2}{3}\right)\theta$$

that is

$$V = 250 \text{ kN} \tag{3.16}$$

(b) Sway Mechanism

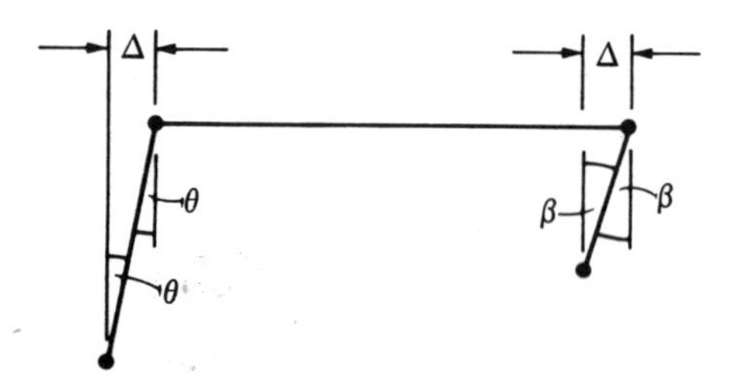

For this mechanism there must be plastic hinges at top and bottom of each column. The geometry requires

$$\Delta = 5\theta = 3\beta$$

that is

$$\beta = \frac{5}{3}\,\theta$$

The work equation is

$$H\Delta = 2 \times 210\theta + 2 \times 210\beta$$

$$H \times 5\theta = 420\left(1 + \frac{5}{3}\right)\theta$$

$$H = 224 \text{ kN} \tag{3.17}$$

(c) Combined Mechanism

The work equation is

$$V \times 4\theta + H \times 5\theta = 600\left(1 + \frac{2}{3}\right)\theta + 420\left(1 + \frac{5}{3}\right)\theta - 2.210\,\theta$$

that is

$$4V + 5H = 1700 \tag{3.18}$$

Equations 3.16 to 3.18 can now be used to draw the ID for this example (figure 3.25). In this case it is natural to use V and H as the axes.

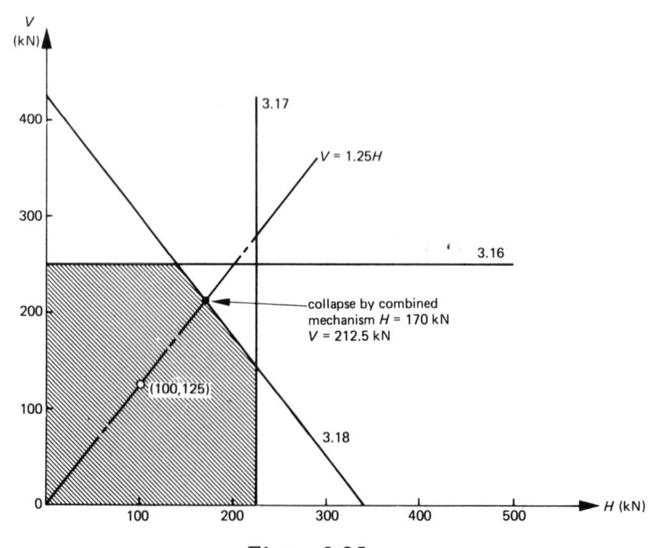

Figure 3.25

What will be the mode of collapse when $V = 1.25H$? Let $H = 100$ kN, then $V = 125$ kN, this is plotted on the ID.

The ID shows that collapse will be by the combined mechanism when $H = 170$ kN, $V = 212.5$ kN. The BMD at collapse for this combination of loads can now be drawn (see appendix C for the method).

The support reactions corresponding to the BMD are also shown (figure 3.26). The vertical reactions are equal to the axial forces in the columns (112.5 kN

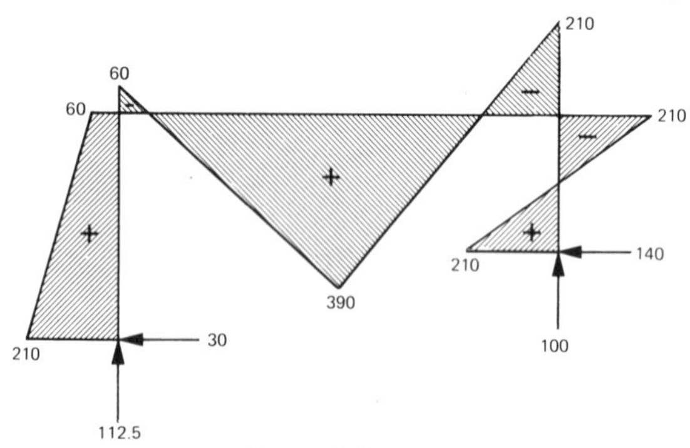

Figure 3.26

and 100 kN) and the difference between the horizontal reactions (110 kN) is the axial force in the beam. As was shown in section 2.5.1 these will reduce the effective plastic moment of each member. The effect of these reductions can be found by repeating the work equations using the smaller plastic moments. The effect is a reduction in the values of H and V at collapse. There would, of course, be a comparable reduction in the member forces ! A cyclic situation is set up, but convergence is rapid. To find the changes in effective plastic moments, information about the member geometries must be available. In this example the plastic moments were based on two universal beam sections, 533 x 165 UB 66 and 305 x 165 UB 54, whose properties are given in reference 2. The modest axial forces reduce the plastic moments to 388 kN m and 208 kN m respectively, and the collapse loads to $H = 168.7$ kN and $V = 210.8$ kN. The reduction in collapse load is only 0.8 per cent. In this case, and for most practical single-storey frames, the axial forces have a negligible effect, but in a multi-storey frame where the axial forces in the lower columns can be very large, the reduction in collapse load can be significant.

3.6.4 Pitched Portal Frame

A typical symmetric pitched portal frame is shown in figure 3.27. This is a very common method of construction for factory or warehouse structures. The rafters allow rainwater to run off, but more importantly they allow large clear spans. The analysis of this type of structure is more complicated than the rectangular portal frame. The complication occurs because the normal beam mechanism cannot develop in the sloping rafters.

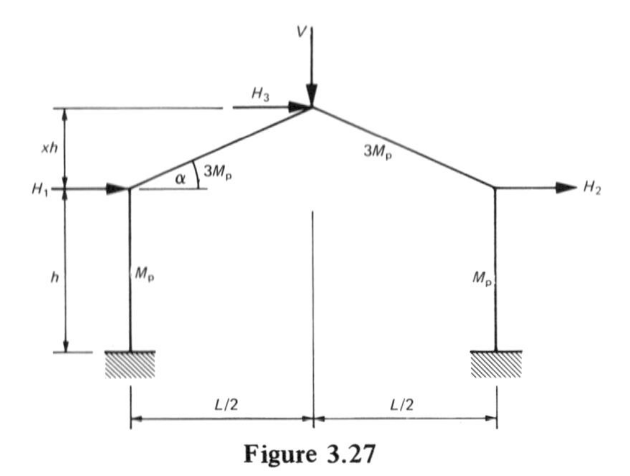

Figure 3.27

Figure 3.28a shows the hinges required for the beam mechanism. However, if the rafter AB rotates about A as in figure 3.28b point B moves vertically and horizontally.

$$\delta_h = AB' \cos(\alpha - \theta) - AB \cos \alpha$$

since $AB' = AB = l$

$$\delta_h = l \cos\alpha \cos\theta + l \sin\alpha \sin\theta - l \cos\alpha$$

For small angles $\cos\theta = 1$, $\sin\theta = \theta$, therefore

$$\delta_h = l \cos\alpha + l (\sin\alpha)\,\theta - l \cos\alpha$$

$$= l (\sin\alpha)\,\theta \qquad (3.19)$$

$$= xh\theta \qquad (3.20)$$

Similarly

$$\delta_v = AB \sin\alpha - AB' \sin(\alpha - \theta)$$

$$= l \sin\alpha - l \sin\alpha \cos\theta + l \cos\alpha \sin\theta$$

$$= l (\cos\alpha)\,\theta = \frac{L}{2}\theta \qquad (3.21)$$

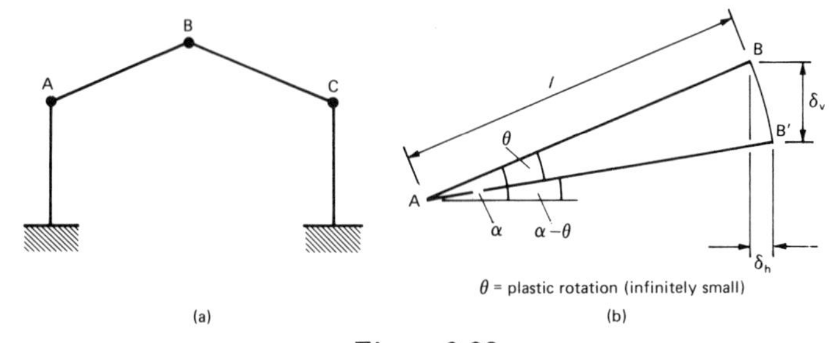

Figure 3.28

Equations 3.20 and 3.21 give the horizontal and vertical deflections of B in terms of the plastic rotation θ

The horizontal deflection is the vertical projection of AB multiplied by the plastic rotation and, similarly, the vertical deflection is the horizontal projection of AB multiplied by the plastic rotation. Hence the vertical deflection must be identical to the deflection of a beam of the same span. There is a similar effect in rafter BC. In order that a beam type of mechanism can occur points A and C must move apart by a distance of $2xh\theta$. The only way that this can happen is for extra hinges to occur in the columns. The various possibilities are shown in figure 3.29. The internal work is the same in each case.

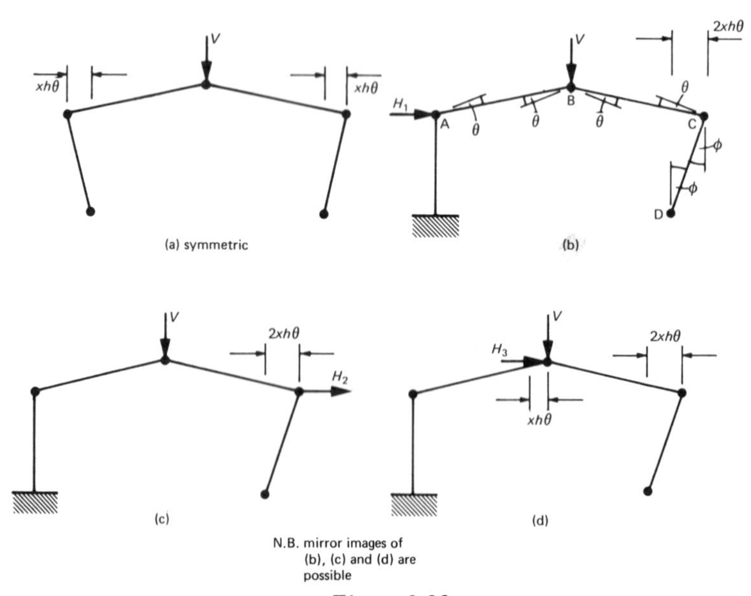

Figure 3.29

Using pattern (b) and the plastic moments in figure 3.29 as an example

$$\text{horizontal movement of C} = 2xh\theta = h\phi$$

that is

$$\phi = 2x\theta$$

$$\text{internal work} = \underset{\text{(hinge at A)}}{M_p\theta} + \underset{\text{(hinge at B)}}{3M_p \times 2\theta} + \underset{\text{(hinge at C)}}{M_p(\theta + \phi)} + \underset{\text{(hinge at D)}}{M_p\phi}$$

$$= 8M_p\theta + 2M_p\phi$$

substituting for ϕ gives

$$\text{internal work} = 4M_p(2 + x)\,\theta$$

(c) and (d) are obviously identical, the reader should check pattern (a). There is another major difference between this and the beam mechanism. This is in the external work, which is different for cases (a) to (d)

cases (a), (b) external work = $VL\theta/2$ (same as beam mechanism)
case (c) external work = $VL\theta/2 + H_2 \times 2xh\theta$
case (d) external work = $VL\theta/2 + H_3 \times xh\theta$

Depending on the position of the horizontal forces, it is possible for both vertical and horizontal forces to do work in this mechanism.

In fact the horizontal forces determine which pattern will occur. Thus (a) can only occur in the absence of horizontal forces (and when the frame itself is completely symmetric). Pattern (b) occurs because H_1 is propping A in position, (c) and (d) because H_2 and H_3 are pulling C and B respectively sideways. There is a more rigorous explanation based on the principle of conservation of energy which the reader may like to work out for himself.

Equation 3.19 relates the horizontal movement in a sloping member to its plastic rotation. In a rectangular frame, angle α is zero (that is, the beam is horizontal) so that there is no horizontal movement. The extra internal work resulting from the spread of the columns in a pitched portal represents extra strength over a rectangular portal of similar span, and accounts for the success of the pitched portal frame.

So far this section has been concerned with symmetric pitched portal frames. Asymmetric frames can be treated in virtually the same way, as is shown in the following example.

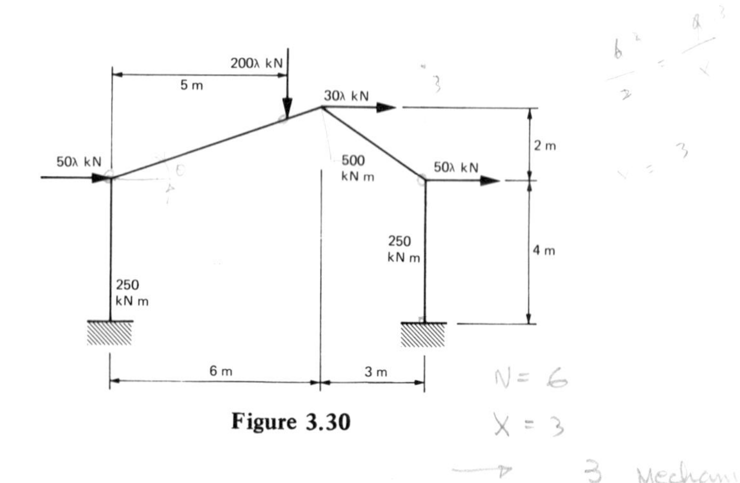

Figure 3.30

The frame to be analysed is shown in figure 3.30. It is required to find the value of the load factor λ when collapse occurs. As before there are three mechanisms to consider, the pitched portal (instead of the beam), the sway, and finally the combined mechanism.

(1) *Pitched Portal Mechanism*

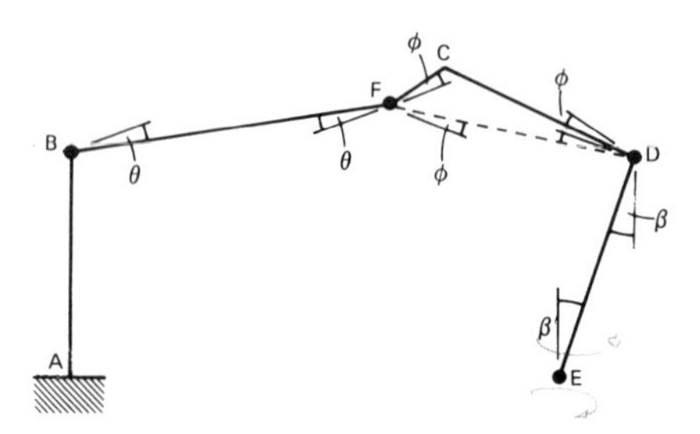

Following the usual rules the hinge positions are as shown. There is apparently a problem because of the kink in FCD. Equations 3.19 to 3.21 were derived for a straight member. However, FCD does not change its shape in any way because all deformation is occurring at the hinges. The movement in FCD is shown in figure 3.31. The horizontal and vertical deflections due to the plastic rotation ϕ at D are

$$\delta_h = FD \sin \alpha\phi \tag{3.22}$$

$$\delta_v = FD \cos \alpha\phi \tag{3.23}$$

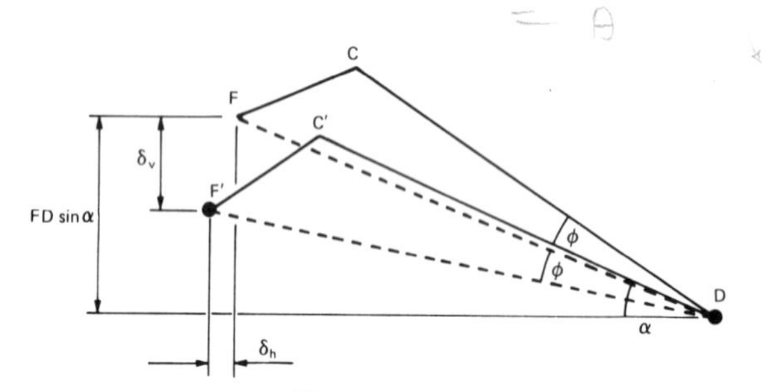

Figure 3.31

which are the same as equations 3.19 and 3.21. Using equations 3.19 to 3.23, the displacements of the loaded points in the frame are

$$\delta_{v_F} = 5\theta = 4\phi$$

$$\delta_{h_F} = \frac{5}{6} \times 2\theta = \frac{5}{3}\theta$$

$$\delta_{h_D} = \frac{5}{3}\theta + \frac{5}{6} \times 2\phi = \frac{5}{3}\theta + \frac{5}{3} \times \frac{5}{4}\theta = \frac{45}{12}\theta$$

$$\delta_{h_C} = \delta_{h_D} - 2\phi = \frac{45}{12}\theta - 2 \times \frac{5}{4}\theta = \frac{5}{4}\theta$$

(Remember that horizontal deflections are the product of the vertical projection of the member and the plastic rotation.) So the work equation is

$$200\lambda\delta_v + 30\lambda\delta_{h_C} + 50\lambda\delta_{h_D} = 250\theta + 500\theta + 500\phi + 250(\phi + \beta) + 250\beta$$

Remembering that δ_{h_D} also equals 4β, ϕ and β can be replaced, giving

$$200\lambda 5\theta + 30\lambda\frac{5}{4}\theta + 50\lambda\frac{45}{12}\theta = 750\theta + 750\frac{5}{4}\theta + 500\frac{45}{48}\theta$$

$$\left(1000 + \frac{150}{4} + \frac{375}{2}\right)\lambda = 750 + \frac{3750}{4} + \frac{22500}{48}$$

$$\frac{9800}{8}\lambda = \frac{17250}{8}$$

$$\lambda = 1.76$$

(2) *Sway Mechanism*

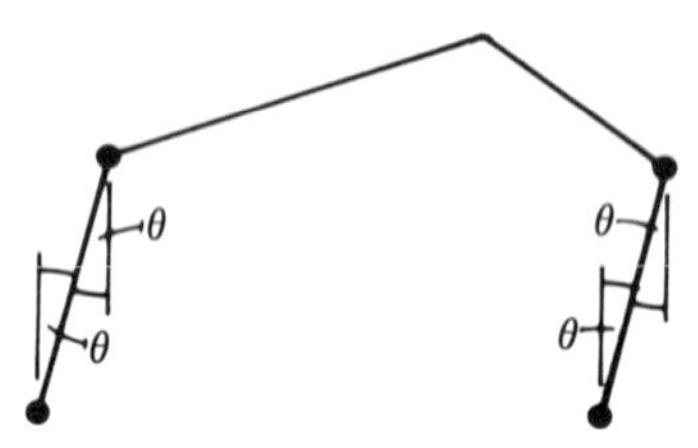

Since deformation only occurs by plastic rotation at the hinges, B, C and D must move sideways by the same amount.

$$(50 + 30 + 50)\lambda 4\theta = 4 \times 250 \times \theta$$

$$520\lambda = 1000$$

$$\lambda = 1.92$$

(3) *Combined Mechanism*

$$\left(\frac{9800}{8} + 520\right)\lambda\theta = \frac{17250}{8}\theta + 1000\theta - 500\theta$$

$$\frac{13960}{8}\lambda = \frac{21250}{8}$$

$$\lambda = 1.52$$

The calculations show that the combined mechanism will occur at the lowest load factor. This mechanism is the critical one for the frame, and the corresponding load factor (1.52 in this case) is called the *collapse load factor.*

Pitched portal frames need a certain amount of care and thought. It is worth while finishing off this section by illustrating this. There is in fact one other likely mechanism in the previous example, a beam mechanism in the left hand rafter.

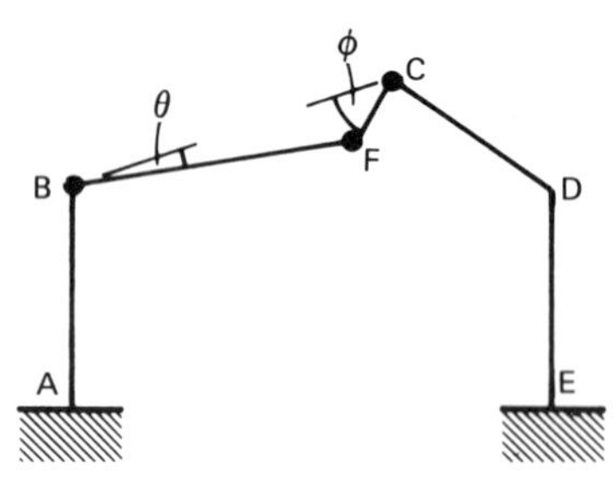

Points B and C *do not* move in this mechanism. BC acts as a beam and point F moves normal to BC. From the geometry of the mechanism

$$BF \times \theta = FC\phi$$

and

$$BF = 5/\cos\alpha$$

$$FC = 1/\cos\alpha$$

so that

$$5\theta = \phi$$

and the work equation is

$$200\lambda 5\theta = 250\theta + 500(\theta + \phi) + 500\phi$$

$$1000\lambda = 250 + 500 \times 6 + 500 \times 5$$

$$\lambda = 5.75$$

Notice in the external work, the distance through which the load moves is 5θ, the vertical movement of F.

It is an unlikely mechanism in a pitched portal frame but it shows what must be done when considering the beam mechanism in a frame such as that shown in figure 3.32.

3.6.5 Summary of the Virtual Work Method

The essential stages of the method are

(1) Identify the possible collapse mechanisms, and impose virtual rotations at the hinges.

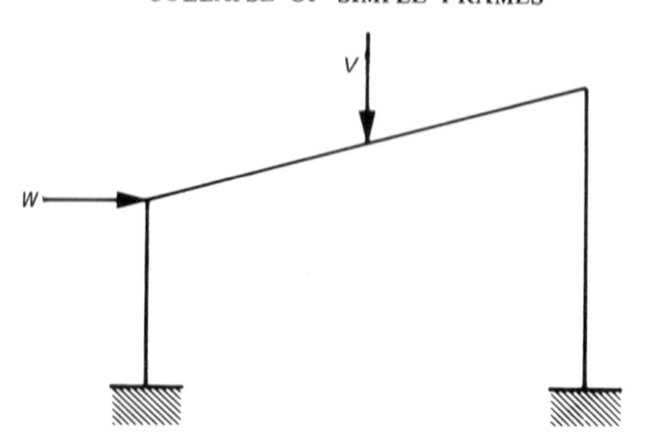

Figure 3.32

(2) For each mechanism, use the geometry of the mechanism to find the relative magnitudes of the plastic rotations and the distances through which the loads move.

(3) Set up the work equation for each mechanism. It should be possible to cancel the magnitude of the virtual rotations.

(4) The final stage depends on what is required of the analysis.

(a) If loads are given in general terms (for example V and H), draw an interaction diagram to describe the collapse behaviour of the structure

(b) If relative magnitudes of the various loads are given, and absolute magnitude determined by a load factor (λ), calculate λ for each mechanism. The critical mechanism has the lowest load factor.

3.7 SUMMARY

The first part of the chapter brought together the material of the first two chapters as applied to framed structures, and provided the essential background to the theorems of plastic analysis. These theorems are the bases for the methods of finding collapse loads. The second part explained in detail the free and reactant BM and virtual work methods. All that remains is to fit both methods into their theoretical context, because that also shows up their limitations.

What has happened in the examples in this chapter? The first thing has been to specify *all* possible collapse mechanisms and then to determine the loads which cause the mechanisms. That is a direct application of the uniqueness theorem. In the final example, the pitched portal frame, the upper bound theorem was also used (unconsciously). The BMDs for each mechanism are shown in figure 3.33. It can be seen that the pitched portal and sway mechanisms do not satisfy the yield condition, confirming that they give upper bounds to the collapse load factor. Only the BMD for the combined mechanism satisfies all three conditions.

The limitation of both methods is that they require every possible mechanism to be examined. There were relatively few possibilities in each example in this

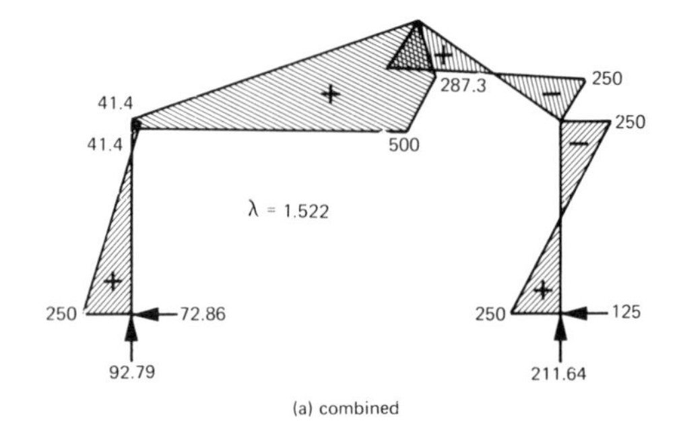

(a) combined

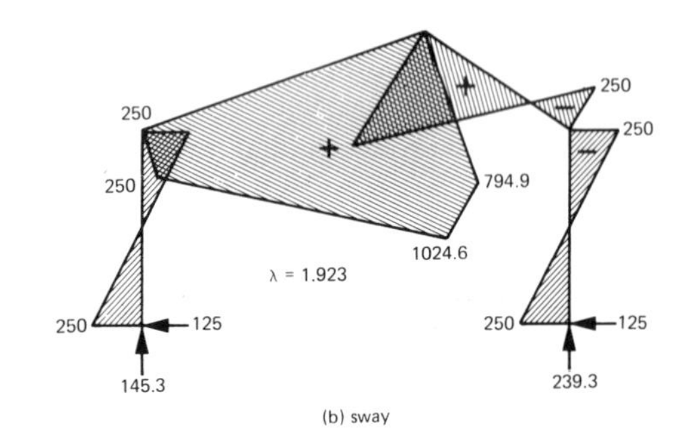

(b) sway

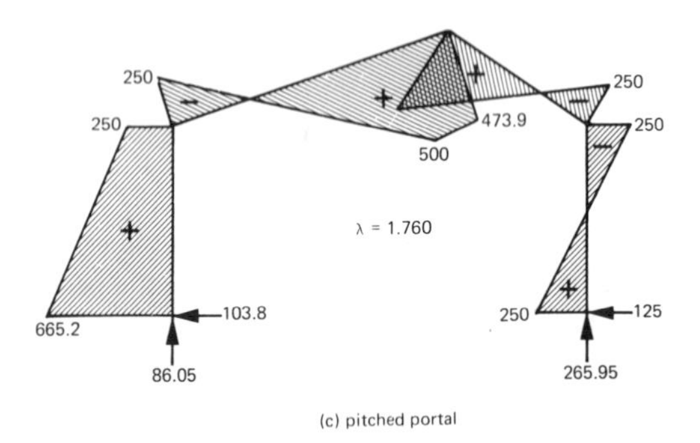

(c) pitched portal

Figure 3.33

chapter so this did not create any difficulty. More complicated frames can have a very large number of possible mechanisms and it would be almost impossible to draw each one, let alone calculate it. Since there is no guarantee that the actual collapse mechanism will be found, the analyses as described cannot be used. Effectively they are limited to beams and single-bay, single-storey frames. There is a method called *limit analysis*, based on the virtual work method, which gets round this problem. It is described in the next chapter.

The other point to note is that, apart from the beam in section 3.5.5, the examples were of structures with concentrated loads. Real structures usually carry distributed loads. Section 3.5.5 showed the problem. The hinge within the span forms at the point of maximum BM, which is not easily defined with a distributed load. Once again it is difficult to determine exactly the collapse mechanism, and limit analysis must be used.

3.8 PROBLEMS

3.1 Calculate the ultimate load of a propped cantilever, span L, carrying a UDL w per unit length. Assume it has a constant plastic moment M_p.

3.2 Calculate the collapse load of the continuous beam shown in figure 3.34. Which is the critical span?

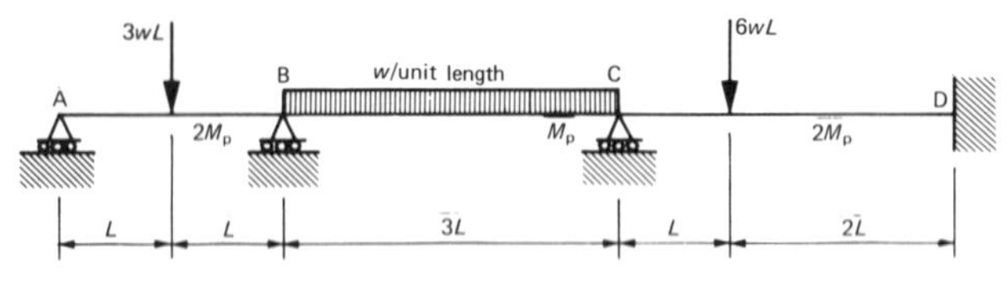

Figure 3.34

3.3 Draw the interaction diagram for collapse of the portal frame shown in figure 3.35. Assume V and H can vary independently of each other.

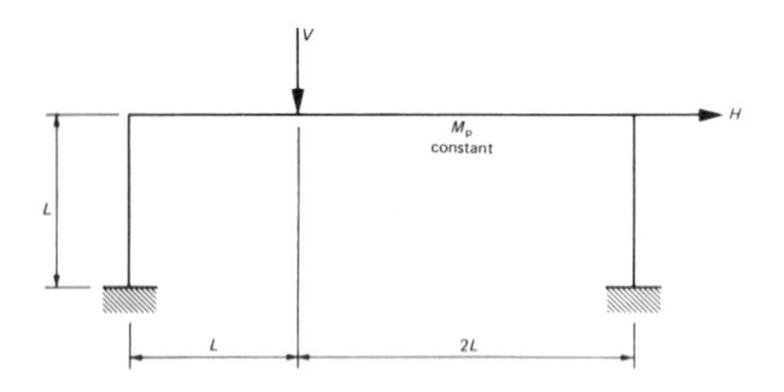

Figure 3.35

3.4 The pinned-foot portal frame in figure 3.36 carries vertical and horizontal forces. If M_p is equal to 300 kNm determine: (a) the critical collapse mechanism, (b) the load factor against collapse, (c) the BMD at collapse.

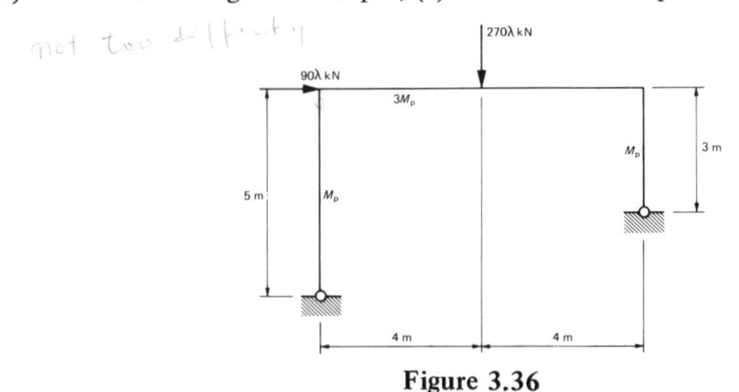

Figure 3.36

3.5 Determine the value of λ at collapse of the frame shown in figure 3.37. Draw the BMD at collapse.

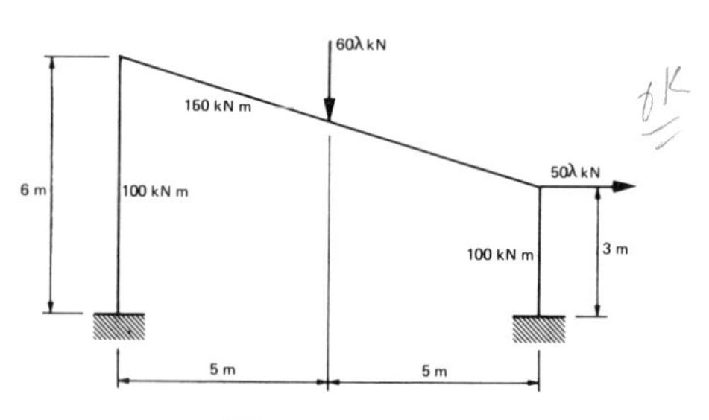

Figure 3.37

3.6 Find the interaction diagram for collapse of the pitched portal frame shown in figure 3.38, assuming V and H can vary independently. What are the collapse loads and mechanisms when $V = H$ and $V = 5H$.

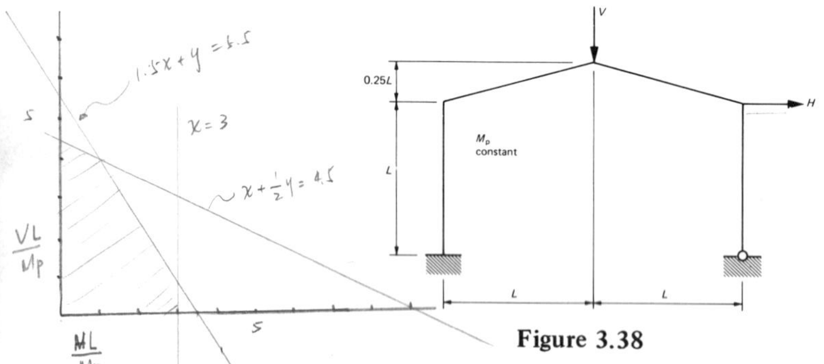

Figure 3.38

3.7 The pitched portal frame shown in figure 3.39 was originally designed to carry a single vertical load $2W$ at the intersection of the rafters BC and CD, the plastic moment M_p for the columns and rafters being equal.

The frame is to be redesigned for the loading shown in the figure. Find the safe value of k if the plastic moment of the columns is now to be reduced to 75per cent of its value in the original design.

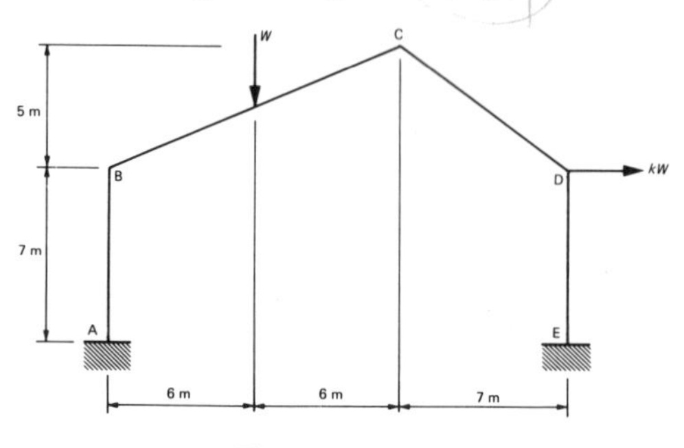

Figure 3.39

4 LIMIT ANALYSIS

4.1 INTRODUCTION

Limit analysis is not related to the limit state philosophy of design. It may seem peculiar to start off this chapter with this statement, but confusion between the two is all too common. Limit state is a philosophy applied to the design of reinforced concrete structures, as embodied in CP 110 [5] and more recently to steel structures, as in the B/20 draft of the Standard for Steelwork Design. [6] Limit analysis is a powerful method for finding the value of, or a range for, the collapse load factor of a structure under a given system of loading, using plastic theory.

The main limitation on the method is that the loading must be proportional, so that the relative magnitudes of the individual loads remain constant and the absolute magnitude is defined by a load factor λ. A separate calculation can be carried out for each combination of loads.

The analysis proceeds as follows.

(1) Guess a collapse mechanism, and determine the load factor λ for the mechanism. In general this mechanism will not be the true collapse mechanism, so from the upper bound theorem

$$\lambda \geqslant \lambda_c$$

(2) Determine the BMD corresponding to λ and the assumed mechanism. If λ is an upper bond, there will be points in the structure where the BMs will be greater than the plastic moment.

(3) Reduce the loads and BMs in the same proportion until all the BMs are less than or equal to the plastic moment. This is achieved by reducing λ to λ_r. At this stage the BM at the original hinge points will be less than the plastic moment, so that the hinges and thus the mechanism no longer exist. This means that at this stage the BMD satisfies the equilibrium and yield conditions but not the mechanism condition. In these circumstances the lower bound theorem states that

$$\lambda_r \leqslant \lambda_c$$

(4) The value of λ_c lies within the range

$$\lambda_r \leqslant \lambda_c \leqslant \lambda$$

It may well be that for practical purposes this range is small enough. If not then a new collapse mechanism can be selected and (1) to (4) repeated.

Step (3) can be confusing. When the original BMs are reduced in the same proportion as the loads they must remain in equilibrium with the loads, and satisfy the yield condition. Hence the new load factor λ_r meets the requirements of the lower bound theorem. It is impossible for the revised BMs to be the same as the BMs which would be found by analysis of the structure with loading defined by λ_r, because the revised BMs contain in some way the redistribution caused by the plastic hinges which would not be present in an exact analysis. This in no way invalidates the procedure. The lower bound theorem only requires a distribution of BMs which satisfies the equilibrium and yield conditions. It does **not** require the BMs to be the actual ones caused by the loads.

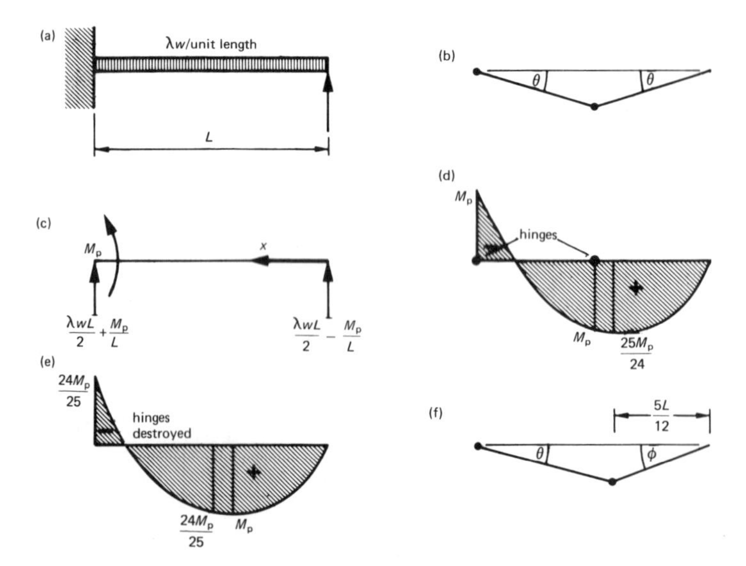

Figure 4.1

The process can be illustrated by means of a propped cantilever with a uniform loading, as in figure 4.1a. Without any prior knowledge of this type of problem a reasonable assumption for the mechanism would be hinges at the fixed support and midspan, as in figure 4.1b. The work equation is

$$\lambda w \times \frac{L}{2} \times \frac{L}{2}\,\theta = 3M_p\theta$$

$$\lambda = \frac{12M_p}{wL^2} \tag{4.1}$$

From the reactions in 4.1c the SF at some general point x is

$$N = -\frac{\lambda wL}{2} + \frac{M_p}{L} + \lambda wx$$

and the SF is zero when

$$\lambda wx = \frac{\lambda wL}{2} - \frac{M_p}{L}$$

substituting equation 4.1 gives

$$\frac{12M_p x}{L^2} = \frac{6M_p}{L} - \frac{M_p}{L}$$

that is

$$x_{crit} = \frac{5L}{12} \tag{4.2}$$

The BM at x is

$$M = \left(\frac{\lambda wL}{2} - \frac{M_p}{L}\right)x - \frac{\lambda wx}{2}$$

and the maximum BM can be found by substituting equations 4.1 and 4.2

$$M_{max} = \left(\frac{6M_p}{L} - \frac{M_p}{L}\right)\frac{5L}{12} - \frac{6M_p}{L^2}\left(\frac{5L}{12}\right)^2$$

$$= \frac{25M_p}{12} - \frac{25M_p}{24}$$

$$= \frac{25M_p}{24}$$

M_{max} can be reduced to M_p by reducing λ to λ_r where

$$\lambda_r = \frac{24}{25} \times \frac{12M_p}{wL^2}$$

$$= \frac{11.52M_p}{wL^2} \tag{4.3}$$

so that

$$\frac{11.52M_p}{wL^2} \leqslant \lambda_c \leqslant \frac{12M_p}{wL^2}$$

This range is small enough, but the process could be continued. An obvious choice of mechanism is hinges at the fixed support and at $5L/12$ from the other support, where the BM was greatest in the previous mechanism. The calculations give

$$\lambda = \frac{11.66M_p}{wL^2}$$

SF = 0 when $x = 0.4142L$

$$M_{max} = M_p \text{ (to 3 decimal places)}$$

$$\lambda_c = \lambda$$

The reader should check these results. As well as illustrating the idea of limit analysis, this example shows that the error involved in assuming a hinge at mid-span with distributed loading is not likely to be great (the error is 3 per cent in the example), and the calculations for the second mechanism are longer and harder for only a minimal increase in accuracy.

It is all very well guessing the collapse mechanism, but in a structure where there may be more than twenty possible mechanisms it is a bit hit and miss. However, the procedure can be given a more rational basis.

4.2 ELEMENTARY MECHANISMS

The examples in chapter 3 contained four types of collapse mechanism: the beam, sway, pitched portal mechanisms and some combination of these. The beam, sway and pitched portal mechanisms with the 'joint rotation' mechanism, are called *elementary mechanisms*. In larger structures any mechanism can be made up by combinations of elementary mechanisms.

The joint rotation mechanism can form at any joint where three or more members meet. The mechanism occurs as in figure 4.2, when plastic hinges form in every member at the joint. Effectively, the joint itself can then rotate with respect to the members. The mechanism can form without any external loading at the joint, so that no work equation can be set up and no corresponding load factor found. It is perhaps a little difficult to envisage at this stage, but its use will become more obvious later.

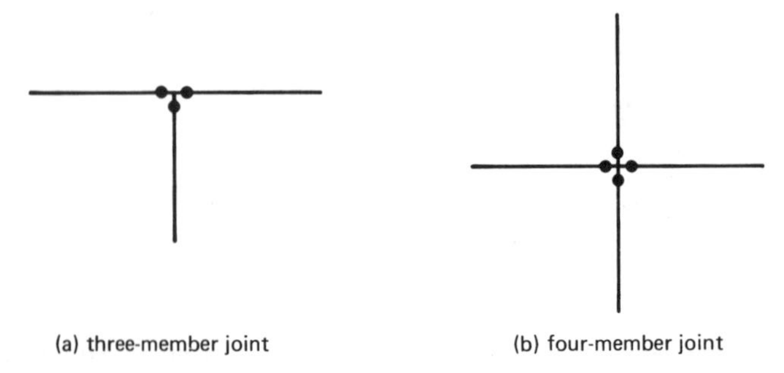

Figure 4.2

The elementary mechanisms are a convenient starting point for the analysis of any frame, and there is a simple test to find the number of elementary mechanisms for the frame. A two-bay frame is shown in figure 4.3 and on it are marked the points at which a plastic hinge could form. In general there will be p such points ($p = 10$ in figure 4.3). The BMD at collapse can be drawn if the BM is

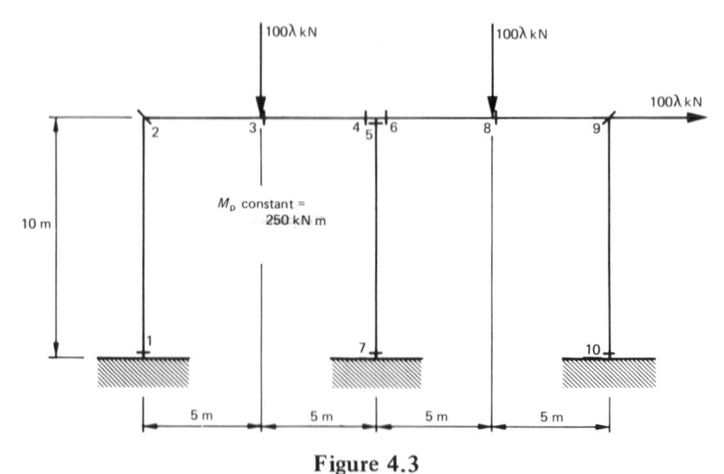

Figure 4.3

known at each of these points, so there are p unknowns. Each of the m elementary mechanisms can be thought of as an independent equation relating moments at the hinge points to the applied loading. Since there are p unknowns and m equations to find them

$$p - m = r$$

where r is the degree of redundancy of the frame. p can be found by a simple count, r by the test in appendix B and the number of elementary mechanisms from the equation

$$m = p - r \tag{4.4}$$

The redundancy test gives $r = 6$ for the frame in figure 4.3, so that $m = 4$. The four mechanisms can be readily identified as two beam mechanisms, a joint rotation and a sway mechanism.

4.3 COMBINATION OF MECHANISMS

The elementary mechanisms represent a series of guesses for the true collapse mechanism, and an estimate for the collapse load factor can be found from each. Other mechanisms can then be created by combining some of the elementary mechanisms. Each new mechanism gives an estimate of the collapse load factor. From the upper bound theorem the lowest estimate will be closest to the actual collapse load factor. There is little point in checking every mechanism by finding the corresponding BMD, but the one with lowest load factor must be checked. The check will either confirm that the actual collapse mechanism and load factor have been found or will provide a range for the collapse load factor.

The combination of mechanisms can be carried out in the same way as in the previous chapter. The individual work equations are added and then the internal work from vanishing plastic hinges subtracted. The following example illustrates the whole process.

4.3.1 Two-bay Frame Example

Figure 4.3 shows a two-bay frame. As was shown earlier it has four elementary mechanisms. The first stage then is to find the load factor for each of these.

Beam mechanisms:

(1) Left-hand beam

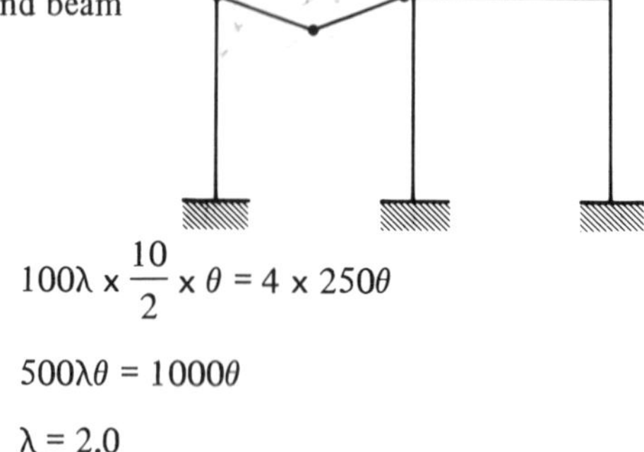

$$100\lambda \times \frac{10}{2} \times \theta = 4 \times 250\theta$$

$$500\lambda\theta = 1000\theta$$

$$\lambda = 2.0$$

(2) Right-hand beam

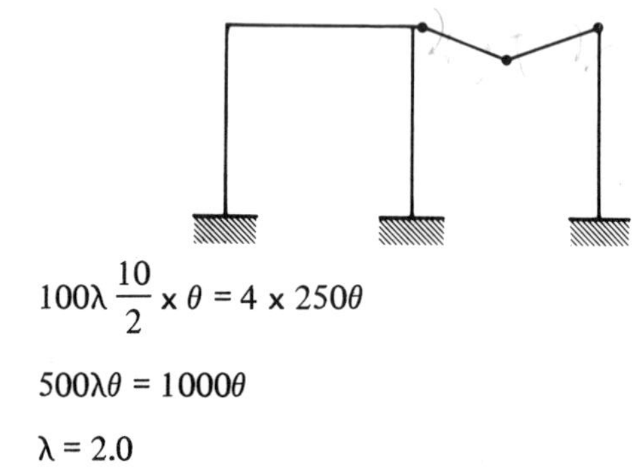

$$100\lambda \frac{10}{2} \times \theta = 4 \times 250\theta$$

$$500\lambda\theta = 1000\theta$$

$$\lambda = 2.0$$

(3) Sway mechanism

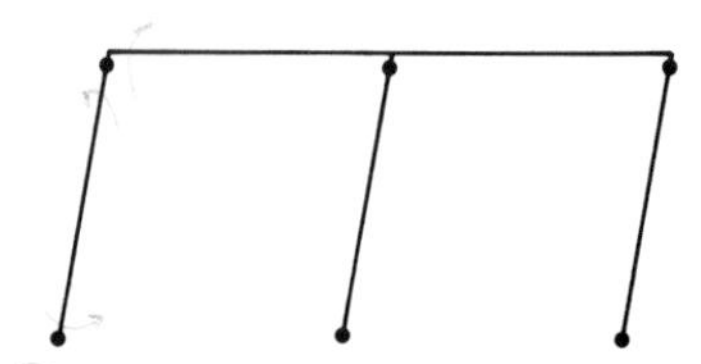

$$100\lambda \times 10 \times \theta = 6 \times 250\theta$$

$$1000\lambda\theta = 1500\theta$$

$$\lambda = 1.5$$

(4) Joint rotation

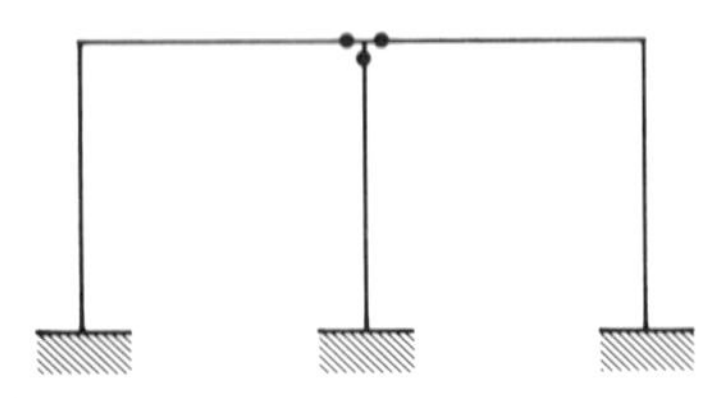

λ indeterminate

The object of the exercise is to find the lowest possible value for λ. The sway mechanism has given the lowest value thus far, so it is the obvious starting point for making any combinations. It is useful to number every mechanism in order to keep track of the calculations.

(5) = (3) + (1). This combination of the left-hand beam and sway mechanisms

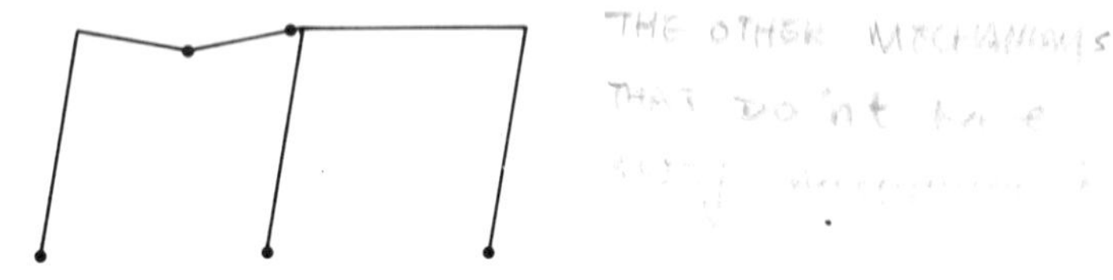

has removed the hinge from the top of the left-hand column. The work equation is

$$1000\lambda\theta + 500\lambda\theta = 1500\theta + 1000\theta - 2 \times 250\theta$$

$$1500\lambda\theta = 2000\theta$$

$$\lambda = 1.33$$

This is an improvement, the load factor is lower than the value for the sway mechanism alone.

(6) = (3) + (2). A combination of the right-hand beam and sway mechanisms. In this case it is not possible to eliminate any hinges because of the relative

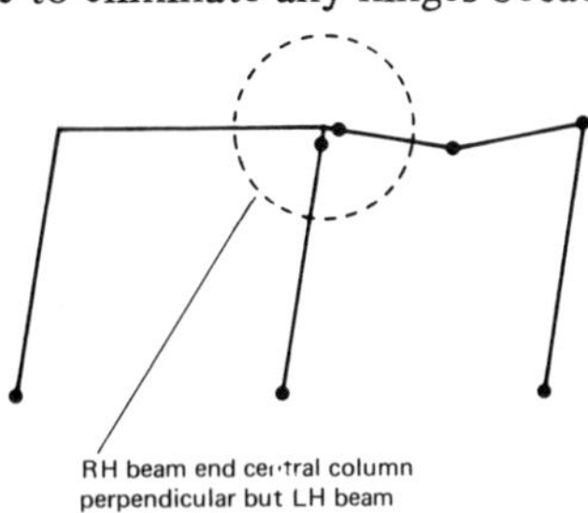

positions of the three members at the central ringed joint.

$$1000\lambda\theta + 500\lambda\theta = 1500\theta + 1000\theta$$

$$1500\lambda\theta = 2500\theta$$

$$\lambda = 1.67$$

This combination is worse. Now consider the ringed joint in more detail. Figure 4.4a shows the joint much enlarged. The joint itself is undeformed, the plastic hinges are just in the right-hand beam and the column. The diagram shows more clearly why two hinges are necessary. The plastic hinges are caused by moments from the loads on the frame, and these moments must act as shown to cause the required rotations. There is a total clockwise moment of 500 kN m.

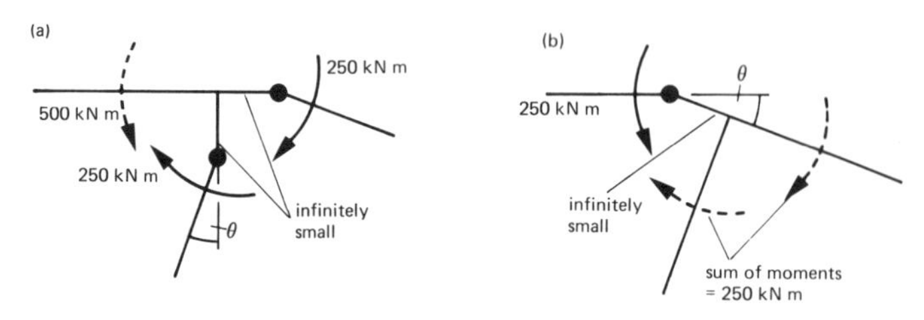

Figure 4.4

To maintain moment equilibrium at the joint there must be an anticlockwise moment of 500 kN m in the left-hand beam. This is physically impossible because the plastic moment of the beam is only 250 kN m, and so the whole joint will rotate. In fact what happens is that the joint rotates at a lower load to the position shown in figure 4.4b. The right-hand beam and the column stay straight and a hinge occurs in the left-hand beam. The overall geometry of the frame remains unaffected by these manoeuvres because the hinges are assumed to form an infinitely small distance away from the joint.

The significance of the joint rotation mechanism now becomes apparent. It describes exactly the change from mechanism (6) to a more realistic situation at the central joint.

(7) = (6) + (4). The joint rotation eliminates hinges in the right-hand beam and central column, but creates a new hinge in the left-hand beam. All this must be reflected in the work equation.

$1500\lambda\theta = 2500\theta - 250\theta \;\; - 250\theta + 250\theta$

no change in external work	RH beam	column	LH beam

$1500\lambda\theta = 2250\theta$

$\lambda = 1.5$

(8) = (7) + (1). This is the only other combination which is possible

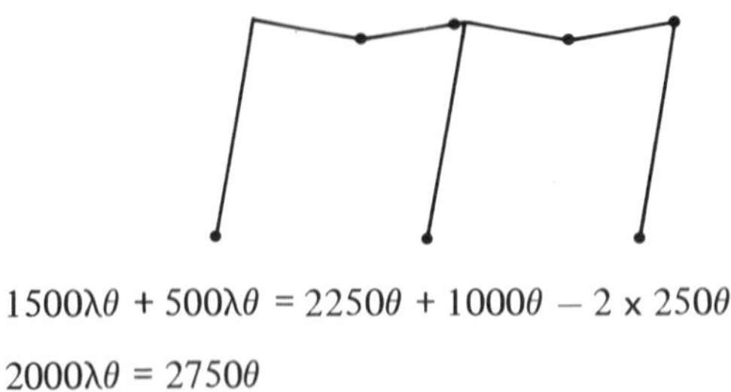

$1500\lambda\theta + 500\lambda\theta = 2250\theta + 1000\theta - 2 \times 250\theta$

$2000\lambda\theta = 2750\theta$

$\lambda = 1.38$

In this example every possible combination has been considered. Mechanism (5) gives the lowest load factor. As a check on the calculations the BMD for the critical mechanism should be drawn. Using the method given in appendix C the diagram in figure 4.5 can be found. The BMD confirms that $\lambda_c = 1.33$ and that mechanism (5) is the actual collapse mechanism, because the BMD satisfies the yield condition (no BM greater than the plastic moment), the mechanism condition (sufficient hinges for a mechanism) and the equilibrium condition.

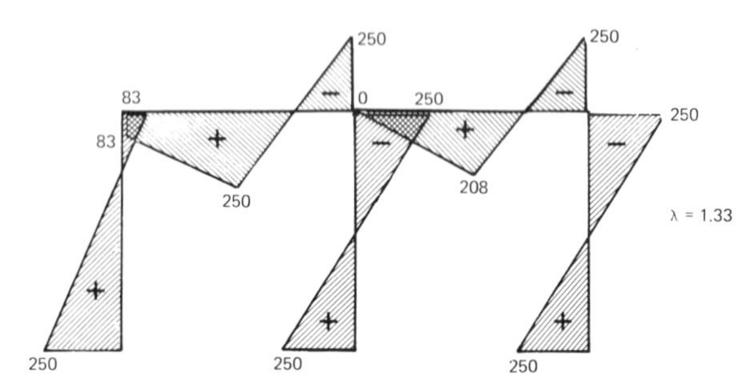

Figure 4.5

Since every possible combination had been tried the exact value (barring errors in the calculation) of λ_c was known in this example. To illustrate what is perhaps the more common situation, assume that mechanism (7) ($\lambda = 1.5$) is critical and check it by finding the corresponding BMD, (figure 4.6).

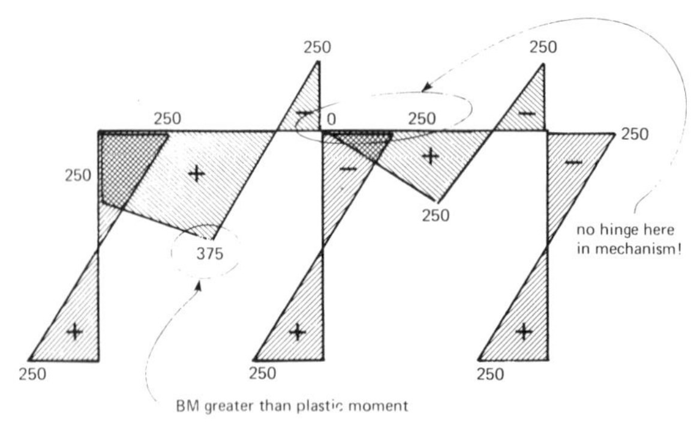

Figure 4.6

The BMD cannot satisfy the yield condition, so $\lambda = 1.5$ must be an upper bound. Reduce everything in proportion until the maximum moment equals the plastic moment. That is

$$\lambda_r = \frac{1.5 \times 250}{375} = 1.0$$

This gives a range for λ_c

$$1.0 \leqslant \lambda_c \leqslant 1.5$$

The average of the limits of the range, 1.25, is only about 6 per cent in error. It must be emphasised that the smaller the range the smaller the error.

In this example there were only four possible combinations and it was practical to try all of them. In more complicated examples this would not be true. Thus it is worth considering a strategy which will reduce the overall effort.

4.3.2 A Strategy for Combining Mechanisms

There is obviously no point in trying to find every possible combination of mechanisms in a complicated structure. What is required is a strategy which will achieve reasonably quickly a close estimate of the collapse load factor.

A general expression for the work equation was given in equation 3.8. This can be rewritten in terms of a load factor as

$$\lambda \Sigma W\delta = \Sigma M_p \theta \tag{4.6}$$

or

$$\lambda = \frac{\Sigma M_p \theta}{\Sigma W\delta} \tag{4.7}$$

When work equations are combined the external work term is always just the sum of the external work from the individual mechanisms. The internal work is found by adding the internal work of the individual mechanisms and then modifying it to account for the hinges which have been eliminated. The object of the combination of mechanisms is to obtain the smallest possible value of λ, and the only way that the value of λ can be affected is by modifying the internal work. The object of any combination of mechanisms must be to *eliminate as many plastic hinges as possible.*

The joint rotation mechanism is very useful for reducing internal work without contributing to the external work. This was shown in mechanism (7) of the previous example, and can be reinforced by another example. A two-storey frame with beam and sway mechanisms is shown in figure 4.7. At the ringed joint the internal work $= 3M_p \times 2\theta = 6M_p\theta$. If a joint rotation is introduced at the joint (notice the rotation of the joint must be 2θ) the internal work at the joint is reduced to $2(M_p 2\theta) = 4M_p\theta$, which gives a 10 per cent reduction in the total internal work.

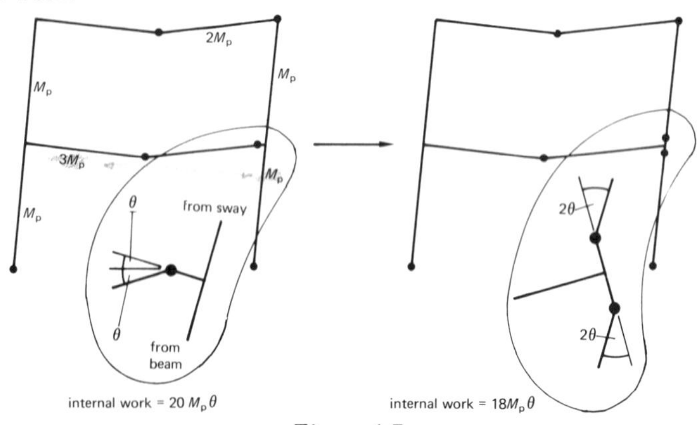

Figure 4.7

The previous paragraph hinted at the major pitfall when combining mechanisms. Remember that hinges are eliminated when the combined mechanisms return parts of the structure to its original geometry, for example a joint returning to a right angle. This can only be done if the rotations in the individual mechanisms are matched correctly, as in figure 4.8.

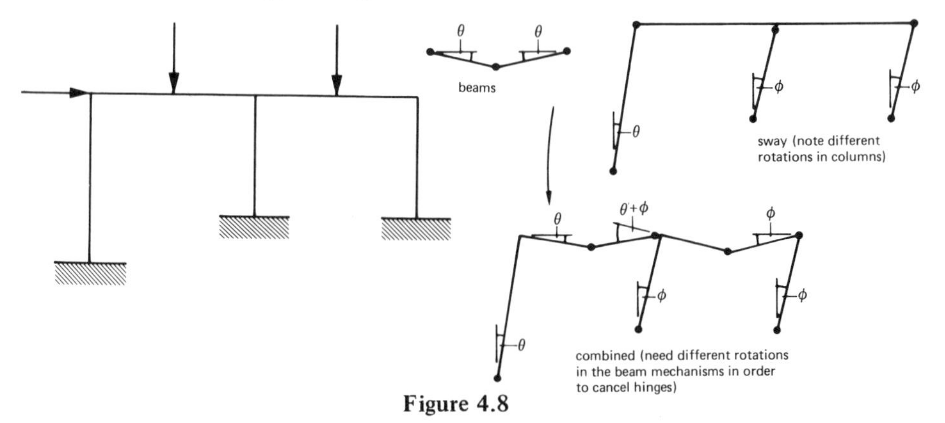

Figure 4.8

4.3.3 Two-Storey Frame with Distributed Loading

Another example of limit analysis is necessary to tie up various ideas which have been brought out in this chapter. The frame shown in figure 4.9 carries distributed loads on the two beams. From the figure it can be seen that

$$p = 12$$

$$r = 6$$

so that

$$m = 6$$

The six elementary mechanisms are two beam, two sway (one for each storey) and two joint rotations. The first stage of the analysis is to write down the work equation for each elementary mechanism.

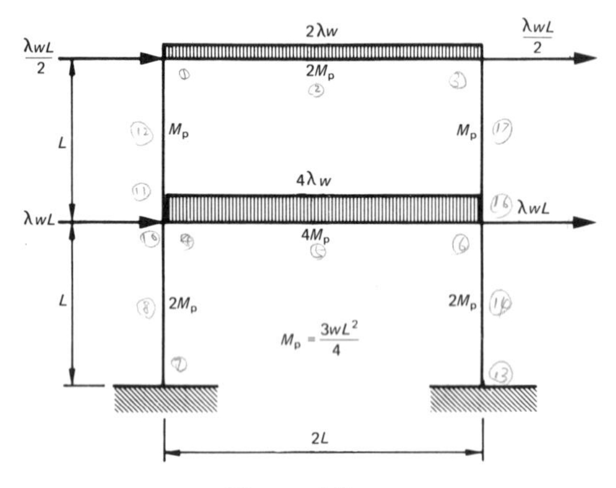

Figure 4.9

Despite the distributed loads on the beams, assume that there are hinges at midspan in the beam mechanisms. This is not correct but, as was shown in section 4.1, the error involved is small. The error can be checked at a later stage.

Diagram of mechanism	Work equation and load factor
1 M_p $2M_p$ M_p upper beam	$$2\lambda wL \times \frac{L\theta}{2} \times 2 = 6M_p\theta$$ $$2\lambda wL^2\theta = 6M_p\theta$$ $$\lambda = \frac{3M_p}{wL^2} = 2.25$$

$- m_1 + 2m_2 - m_3 = 2 \cdot \lambda \cdot \frac{4}{3} M_p$

$\rightarrow m_1 = -(4.8 - 5) = -0.2$

Diagram of mechanism — **Work equation and load factor**

2

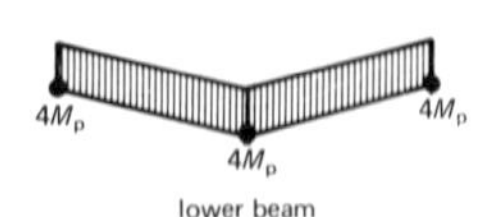

lower beam

$$4\lambda wL \times \frac{L\theta}{2} \times 2 = 16M_p\theta$$

$$4\lambda wL^2\theta = 16M_p\theta$$

$$\lambda = \frac{4M_p}{wL^2} = 3.0$$

3

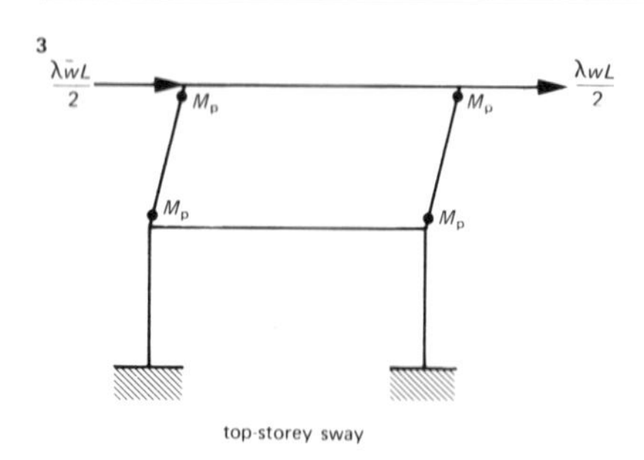

top-storey sway

$$\frac{\lambda wL}{2} \times L\theta \times 2 = 4M_p\theta$$

$$\lambda wL^2\theta = 4M_p\theta$$

$$\lambda = \frac{4M_p}{wL^2} = 3.0$$

4

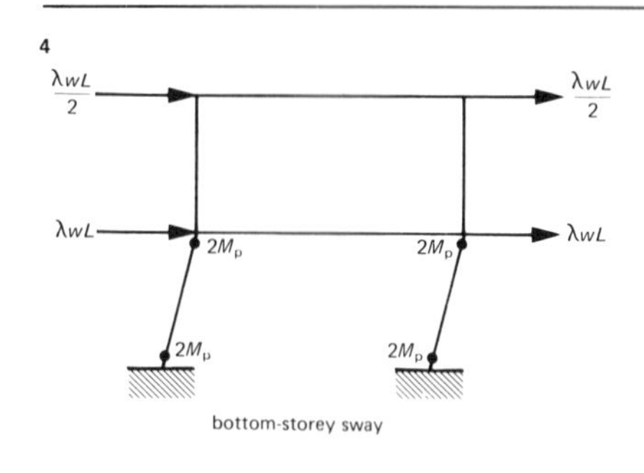

bottom-storey sway

Note: all horizontal loads move in this mechanism, and must be in the work equation.

$$\lambda wL \times L\theta \times 2 + \frac{\lambda wL}{2} \times L\theta \times 2 = 8M_p\theta$$

$$3\lambda wL^2\theta = 8M_p\theta$$

$$\lambda = \frac{8M_p}{3wL^2} = 2.0$$

5

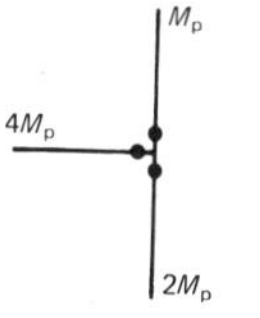

left-hand joint rotation

No work equation

λ indeterminate

6

M_p

4M_p

2M_p

right-hand joint rotation

No work equation

λ indeterminate

Diagram of mechanism	Work equation and load factor

Bottom-storey sway has lowest load factor, obvious starting point for combinations.

---

7 = 4 + 2 + 5 + 6

Joint rotations included immediately to reduce internal work

$$3\lambda wL^2\theta + 4\lambda wL^2\theta =$$

$$8M_p\theta + 16M_p\theta - 6M_p\theta + M_p\theta - 4M_p\theta + 3M_p\theta$$

$$7\lambda wL^2\theta = 18\, M_p\theta$$

$$\lambda = \frac{18}{7}\,\frac{M_p}{wL^2} = 1.929$$

---

8 = 4 + 3

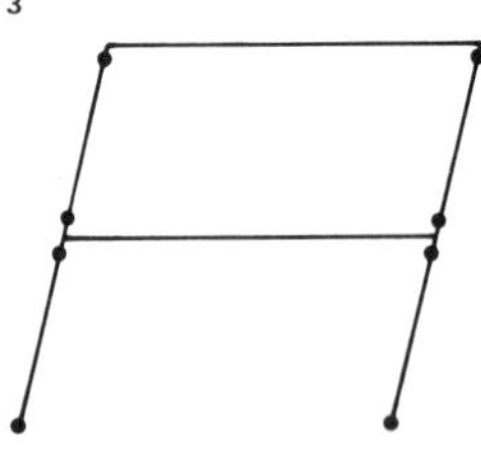

Only an intermediate step included for clarity. No hinges are eliminated by this combination.

$$3\lambda wL^2\theta + \lambda wL^2\theta = 8M_p\theta + 4M_p\theta$$

$$4\lambda wL^2\theta = 12M_p\theta$$

$$\lambda = \frac{3M_p}{wL^2} = 2.25$$

---

9 = 8 + 1 + 2 + 6

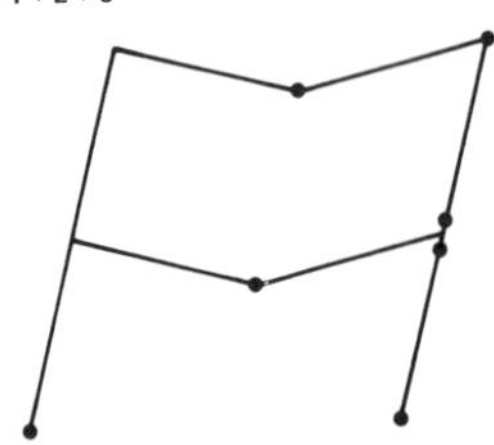

$4\lambda wL^2\theta + 2\lambda wL^2\theta + 4\lambda wL^2\theta = 12M_p\theta + 6M_p\theta$
$+ 16M_p\theta - 2M_p\theta$ (top LH joint) $- 7M_p\theta$
(lower LH joint) $- 4M_p\theta + 3M_p\theta$

$$10\lambda wL^2\theta = 24M_p\theta$$

$$\lambda = \frac{24}{10}\,\frac{M_p}{wL^2} = 1.8$$

This is the lowest load factor, check this mechanism by finding the BMD

The BMD for mechanism (9) can be found using the method in appendix C. (In fact this example is used in the appendix.) For this type of frame it is necessary to consider the free and reactant BMs in each beam, and horizontal equilibrium for each storey separately.. The final BMD is shown in figure 4.10

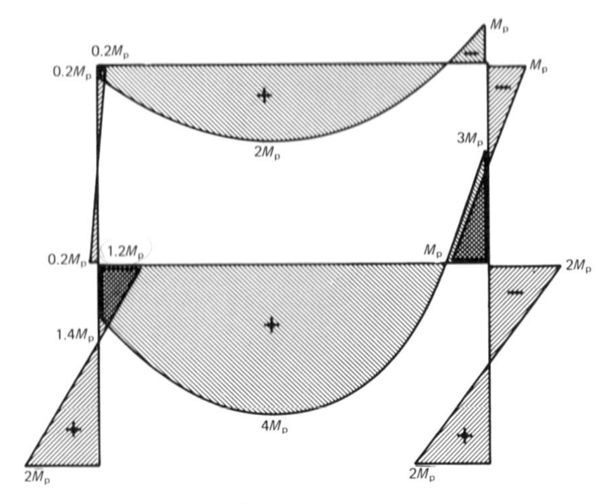

Figure 4.10

and apparently it satisfies all the conditions. The conclusion would be that this is the correct mechanism, but because of the distributed loading, the values of the moments in each beam must be checked.

(a) *Upper Beam*

The upper beam is shown in figure 4.11 removed from the rest of the frame. The end moments are known from the BMD, the unknown reactions at A and B are the axial forces in the two upper columns. Taking moments about B gives

$$V_{\mathrm{A}} \times 2L + 0.2M_{\mathrm{p}} + M_{\mathrm{p}} - 3.6w \times 2L \times L = 0$$

$$V_{\mathrm{A}} = \frac{7.2wL^2 - 1.2M_{\mathrm{p}}}{2L} = \frac{7.2wL^2 - 1.2 \times 0.75wL^2}{2L}$$

$$= 3.15wL$$

The BM (sagging positive) at any point along the beam is thus

$$M = 0.2M_{\mathrm{p}} + 3.15wLx - \frac{3.6wx^2}{2}$$

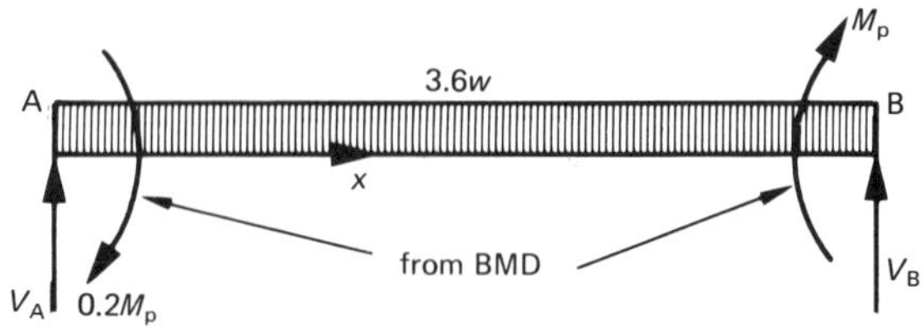

A 3.6w B M_{p} x from BMD V_{A} $0.2M_{\mathrm{p}}$ V_{B}

Figure 4.11

ok

$dM/dx = 3.15wL - 3.6wx = 0$ for maximum moment. That is $x = 3.15L/3.6 = 0.875L$ at the point of maximum moment

$$M_{max} = 0.2M_p + 3.15wL(0.875L) - \frac{3.6w(0.875L)^2}{2}$$

$$= 0.2M_p + 1.378wL^2$$

$$= 2.04M_p$$

(b) *Lower Beam*

The beam is shown in figure 4.12. The calculations are very similar to those for the upper beam.

$$V_c \times 2L + 1.4M_p + 3M_p - 7.2w \times 2L \times L = 0$$

$$V_c = 5.55wL$$

$$M = 1.4M_p + 5.55wLx - \frac{7.2wx^2}{2}$$

$$\frac{dM}{dx} = 5.55wL - 7.2wx$$

$x = 0.771L$ at the point of maximum moment

$$M_{max} = 4.25M_p$$

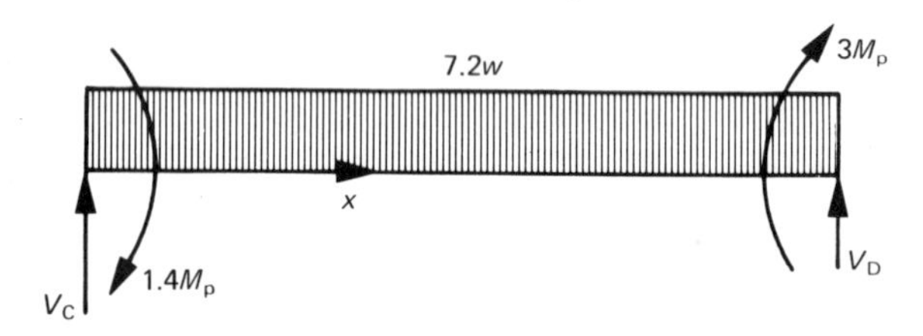

Figure 4.12

There are moments in both beams which are greater than the plastic moment of the section. This must mean that the BMD fails the yield condition and the calculated load factor must be an upper bound on λ_c. A lower bound can be found by reducing everything in proportion until the maximum moment equals the plastic moment. Thus

$$\lambda_r = \frac{2 \times 1.8}{2.04} = 1.76$$

or

$$\lambda_r = \frac{4 \times 1.8}{4.25} = 1.69$$

The range for the collapse load factor is

$$1.69 \leqslant \lambda_c \leqslant 1.8$$

with the average of the limits 1.75. These limits are close enough for any practical problem.

However, it is worth going on in this case to show the effect of distributed loading in a more complex structure. Mechanism (9) is obviously not correct, but until the BM distribution in each beam was checked, it did appear to be satisfactory. The inference from this is that the true collapse mechanism will be very similar, but with hinges near to the points of maximum moment, rather than at midspan, as in figure 4.13

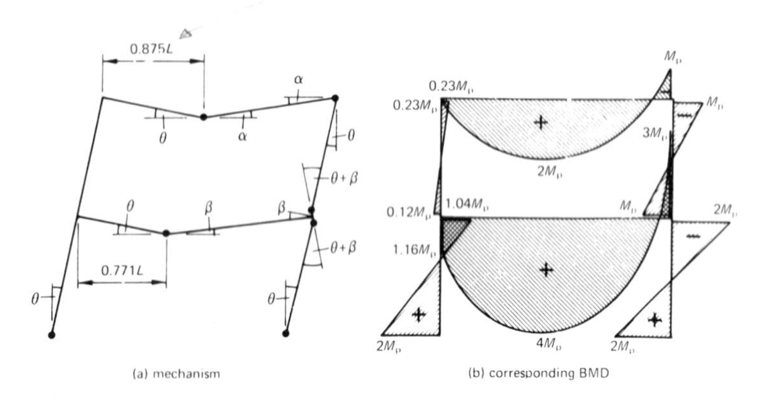

Figure 4.13

The work equation for this mechanism could be found by combination of mechanisms, after rewriting the beam mechanisms, or as follows, by using the rotations in the figure. From the mechanism geometry

$$1.125\alpha = 0.875\theta$$

$$\alpha = 0.778\theta$$

$$1.299\beta = 0.771\theta$$

$$\beta = 0.627\theta$$

Starting at the bottom of the frame and working upwards

$$\begin{aligned}\text{internal work} &= 2M_p\theta + 2M_p\theta + 4M_p(\theta + \beta) + 2M_p(\theta + \beta) + M_p(\theta + \beta) + \\ &\quad 2M_p(\theta + \alpha) + M_p(\theta + \alpha) \\ &= 14M_p\theta + 7M_p\beta + 3M_p\alpha \\ &= 20.73M_p\theta\end{aligned}$$

$$\text{External work} = 4\lambda wL^2\theta + \frac{2\lambda w(0.875L)^2\theta}{2} + \frac{2\lambda w(1.125L)^2\alpha}{2} + \frac{4\lambda w(0.771L)^2\theta}{2}$$

horizontal loads — top beam — bottom beam

$$+ \frac{4\lambda w(1.229L)^2\beta}{2}$$

$$= 5.9545\lambda wL^2\theta + 1.2656\lambda wL^2\alpha + 3.0209\lambda wL^2\beta$$

$$= 8.834\lambda wL^2\theta$$

Equating internal and external work gives

$$8.834\lambda wL^2\theta = 20.73M_p\theta$$

$$\lambda = \frac{20.73M_p}{8.834wL^2} = 1.76$$

The BMD for the mechanism is shown in figure 4.13b. The BM distributions in the beams show that

(1) for the upper beam $x = 0.869L$ at the point of maximum moment

$M_{max} = 2M_p$ (actually it is $2.00002M_p$)

(2) for the lower beam $x = 0.779L$ at the point of maximum moment

$M_{max} = 4M_p$ (actually $4.0002M_p$)

The reader might like to check these results. The hinges are not in exactly the correct positions, but the results show that effectively the mechanism is the correct one. The collapse load factor is 1.76, but the error from assuming the hinges are at midspan is only 2.3 per cent. The error is so small that it would be difficult to justify the extra effort of considering the more accurate mechanism.

There was another saving of effort during the combination process. The joint rotations were combined at the same time as the beam mechanisms to reduce the internal work. The ability to spot such savings comes with experience.

4.4 SUMMARY

The chapter has dealt with the *limit analysis* of framed structures carrying *proportional* loading as embodied in the *combination of mechanisms* method. The result of the method is either the value of, or a range for, the *collapse load factor* of the frame.

The main stages in the analysis are

(1) Identify the elementary mechanisms and the corresponding work equations.

(2) Starting with the elementary mechanism with the lowest load factor, combine mechanisms to cancel plastic hinges by maintaining locally the original geometry of the structure (keep members straight or joints perpendicular, for example). Find the load factor of each combined mechanism.

(3) The lowest load factor is closest to the collapse load factor (from the upper bound theorem). Check the mechanism by finding the corresponding BMD.

(4) If the BMD satisfies the yield, mechanism and equilibrium conditions, the mechanism is the actual collapse mechanism of the frame. The load factor is equal to the collapse load factor.

(5) If the BM is greater than the corresponding plastic moment somewhere in the structure, the mechanism is not the correct one and the load factor λ is greater than λ_c. Reduce the value of λ and all the BMs in the same proportion until all BMs are less than or equal to the corresponding plastic moment. From the lower bound theorem the reduced load factor λ_r is less than λ_c, so that

$$\lambda_r < \lambda_c < \lambda$$

It is usually sufficiently accurate to assume that a plastic hinge forms at midspan in a member carrying a distributed load, although the result will be an overestimate of λ_c.

The best way to become familiar with this method is by practice. The examples provided bring out the various points made in this chapter: do try them.

4.5 PROBLEMS

4.1 Find the collapse load factor of the frames shown in figures 4.14a to 4.14e. In each case check the critical mechanism by drawing the corresponding BMD.

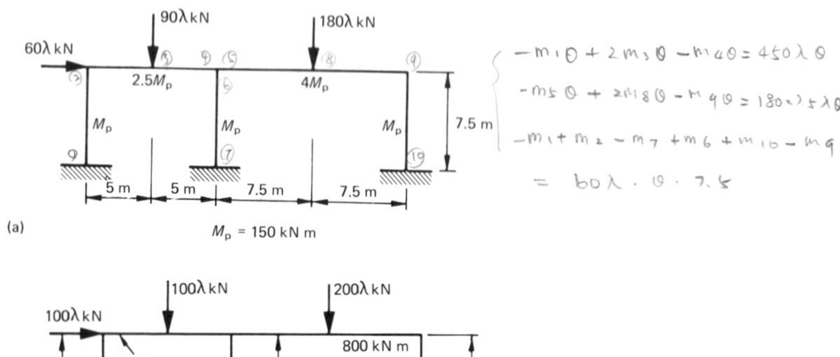

Figure 4.14a-b

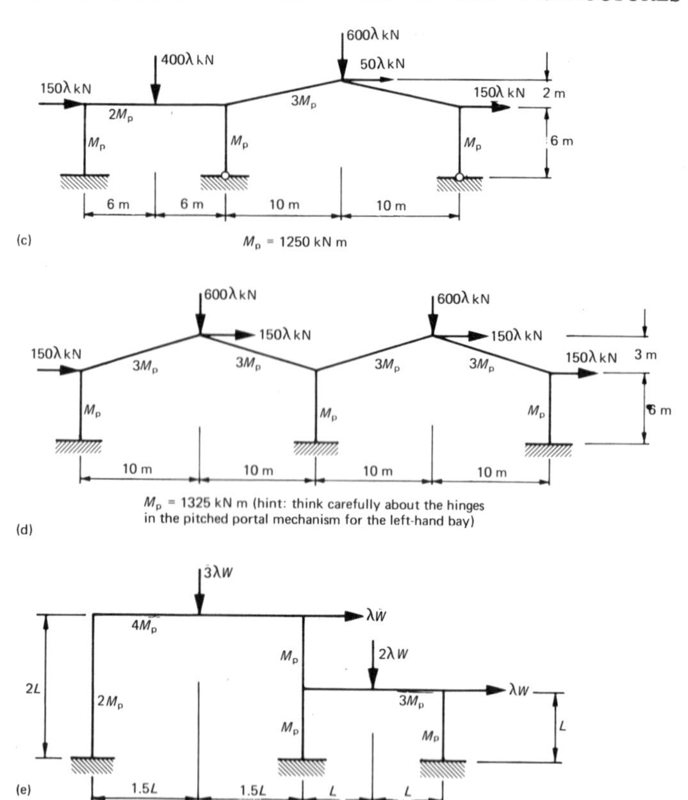

Figure 4.14c-e

4.2 Find the collapse load factors of the frames shown in figures 4.15a and 4.15b, assuming that plastic hinges form at midspan. Because of the UDLs these must be upper bounds; use limit analysis to find a range for the actual collapse load factor.

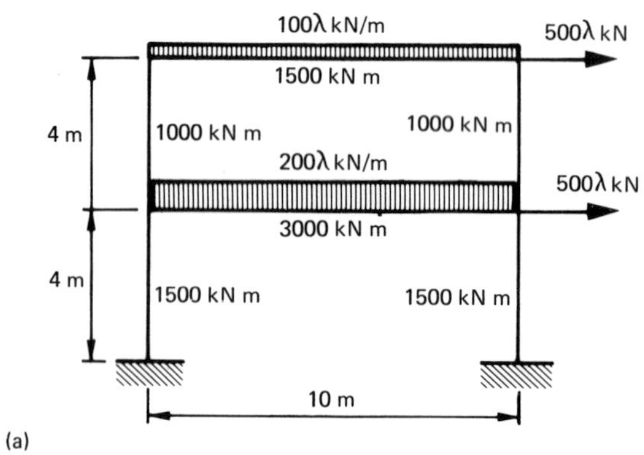

Figure 4.15a

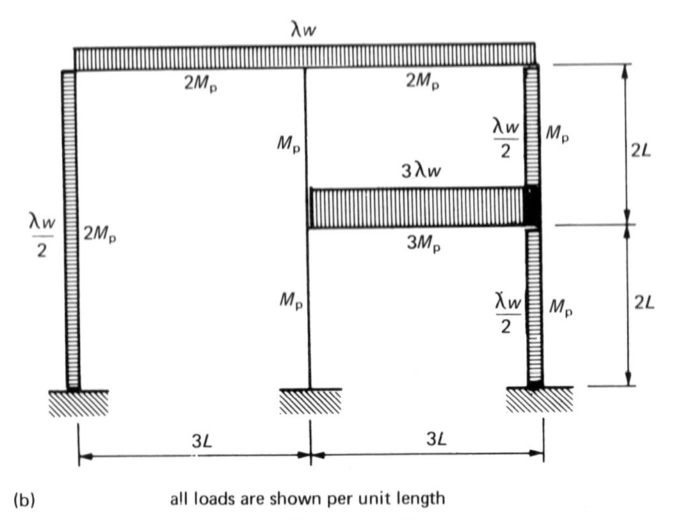

(b) all loads are shown per unit length

Figure 4.15b

5 DESIGN USING PLASTIC THEORY

5.1 INTRODUCTION

The last two chapters have been concerned with *analysis*, that is, solving problems where the basic geometry of the structure (length of beams and column heights) and the size of the members is given. This would be the situation when an existing structure is checked for resistance to collapse. More frequently the problem is to find the size of the members, given the loading (or loadings) and the basic geometry, such that the structure has a required load factor against collapse. The sizing of the members is a *design* problem.

Design is in some ways the reverse of analysis, and can still be carried out using plastic theory. Before looking at plastic design it is worth while to consider the question 'why bother with plastic design?'. There are four main reasons why.

(1) In the United Kingdom the trend is to adopt the limit state philosophy of design. This has already been achieved in CP 110 [5] for Concrete Structures and is embodied in the B/20 Draft Specification for Steelwork [6] (which is intended as a replacement for BS 449. [7]) Most frequently it is the ultimate limit state (the resistance to collapse) which is critical in defining the overall design. Obviously, plastic theory is directly applicable to the ultimate limit state. The current American specification for Structural Steel [38] was adopted in 1969. It is based on what is called the *working stress philosophy of design*. However it does allow design by plastic methods.

(2) As has been shown already, the calculations required by plastic theory are easier than those required by elastic theory, and no more tedious. Easier calculations must lead to quicker and more reliable results. 'Lack of fit' in the members does not affect collapse (see section 3.2.3).

(4) Plastic methods give the designer considerable flexibility. It is the designer who is in control, rather than the member properties. This will be shown later in this chapter and to a greater extent in chapter 8 when the design of reinforced concrete slabs is considered.

The picture is not entirely rosy, of course. There are two reservations which must be borne in mind.

(1) There are designs in which the ultimate limit state is not critical. For example, in some structures strict limits on deflections at normal working loads (when the structure is probably elastic) are more critical than resistance to collapse.

(2) The plastic methods make no allowance for the possibility of buckling. Buckling problems are always possible, and any satisfactory design must eliminate them. The effects of buckling are considered in the next chapter.

5.2 LOAD FACTORS

The object of any design based on plastic theory must be to produce a structure with a specified load factor against collapse. The value of the load factor to be used is in itself a complex subject, open at the moment to debate due to the impending release of a new British Standard for steelwork design. This debate is outside the scope of this book because it has nothing to do with plastic theory. However, it is worth while presenting the existing and proposed load factors.

The current British Standard, BS 449: 1969, allows design by plastic methods without stating explicitly the required load factor. BS 449 is based mainly on elastic theory and it can be shown that the load factor for a single-span beam resulting from the elastic requirements varies with the end support conditions and type of loading. The minimum value is 1.75 for a simply supported I-beam. It was argued [8] that this value could be adopted for any structure. BS 449 recognises that it is highly unlikely that maximum wind load and maximum imposed load will occur simultaneously so that the load factor for such a load combination could be reduced. Consequently the commonly accepted load factors for design to BS 449 are

Imposed and dead load – 1.75

Wind and imposed and dead load – 1.4

Part of the limit state philosophy adopted in the B/20 draft of the proposed new standard is to base load factors on the probability of occurrence of various load combinations. Consequently different collapse load factors are proposed for various loadings as shown in table 5.1. There is a further load factor in every case to allow for the possibility of variation in strength of the steel. This is done by using an effective yield stress equal to $\sigma_y/1.075$. This increases the values in table 5.1 by 7.5 per cent. Even so, the proposed load factors are lower than those in BS 449, reflecting the increase in confidence in the plastic methods of design.

The American specification for Structural Steel [38] adopts a similar, but more detailed approach than BS 449. It allows plastic methods of design and requires a collapse load factor of 1.7 on live and dead loads, and a reduced factor of 1.3 when these loads act in conjunction with wind or earthquake loads.

The various examples in this chapter use arbitrarily chosen load factors to illustrate the principles of the plastic methods. In one or two cases the values do have a striking similarity to the BS 449 values.

Table 5.1

Type of load or combination of loads		γ_f Factor
Dead load	Maximum	1.4
	Minimum	1.0
	Minimum for pattern loading	1.2
Imposed load (in the absence of wind load)		1.6
Wind load (acting with dead load only)		1.4
Wind and imposed loads (acting in combination)		1.2

5.3 EXAMPLE OF DESIGN BY PLASTIC THEORY

The beam shown in figure 5.1 is to be designed by plastic methods. The example is rather idealised because only one combination of loads is considered (incidentally the loads include an allowance for the self-weight of the beam) but it does illustrate the general approach to plastic design. Initially it is assumed that each span has a different size member. The first stage is to examine all the

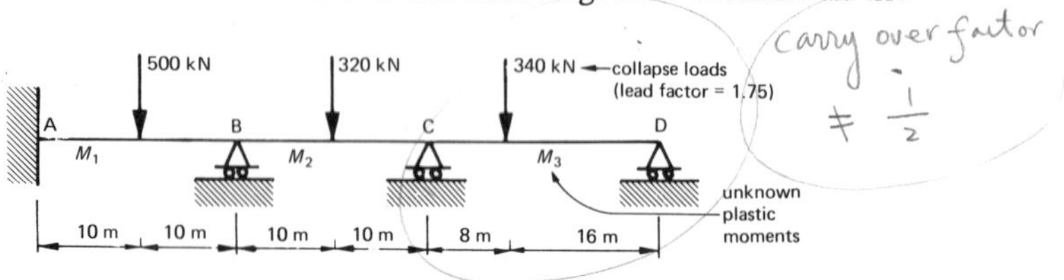

Figure 5.1

possible collapse mechanisms, which in this example are beam mechanisms in each span. Simultaneous collapse of the spans will produce the most economic design, and this can be achieved by choosing appropriate member sizes. In this circumstance the BMD at collapse will be as in figure 5.2. The free and reactant BMDs can be used to find the actual BM at any point.

$$\text{span AB: free BM}_{max} = \frac{500 \times 20}{4} = 2500 \text{ kN m}$$

$$\text{span BC: free BM}_{max} = \frac{320 \times 20}{4} = 1320 \text{ kN m}$$

$$\text{span CD: free BM}_{max} = \frac{340 \times 8 \times 16}{24} = 1813 \text{ kN m}$$

There is no difficulty with the free BMs but consider the reactant BM at the

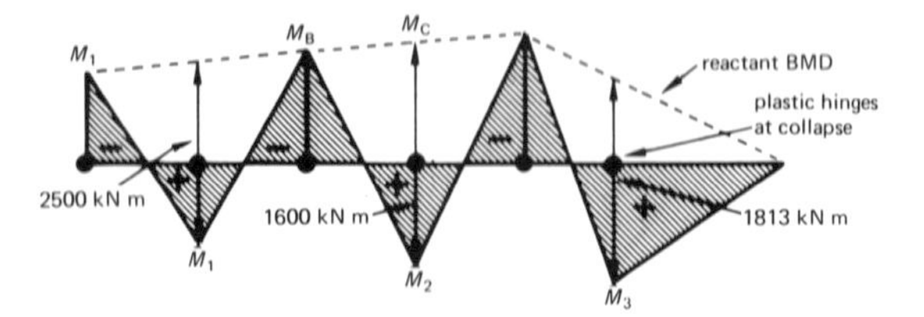

Figure 5.2

supports B and C where there are plastic hinges. The value of the BM is shown as M_B and M_C at B and C respectively. The moments are equal to the smaller plastic moment of the two members meeting at B and C, but the relative magnitudes of the plastic moments are unknown at this stage

$$M_B = \text{smaller of } M_1 \text{ and } M_2$$

$$M_C = \text{smaller of } M_2 \text{ and } M_3$$

This impasse can be removed by considering each span in turn and making some guesses.

Span AB: guess that $M_1 < M_2$. Figure 5.3 shows the corresponding BMD. The geometry of the BMD requires that

$$2M_1 = 2500 \text{ kN m}$$

$$M_1 = 1250 \text{ kN m} \tag{5.1}$$

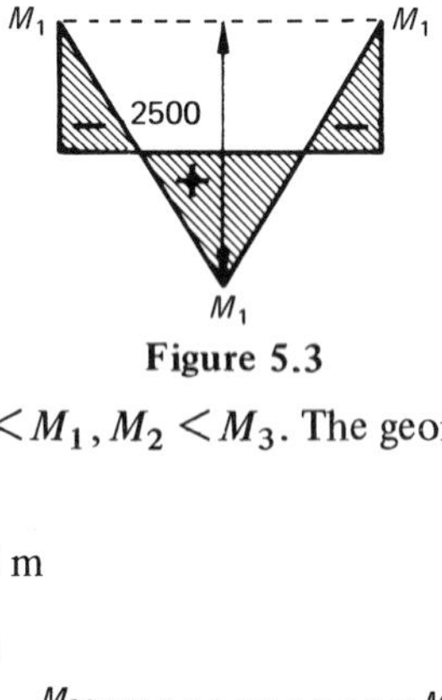

Figure 5.3

Span BC: guess that $M_2 < M_1, M_2 < M_3$. The geometry of the BMD in figure 5.4 requires that

$$2M_2 = 1600 \text{ kN m}$$

$$M_2 = 800 \text{ kN m} \tag{5.2}$$

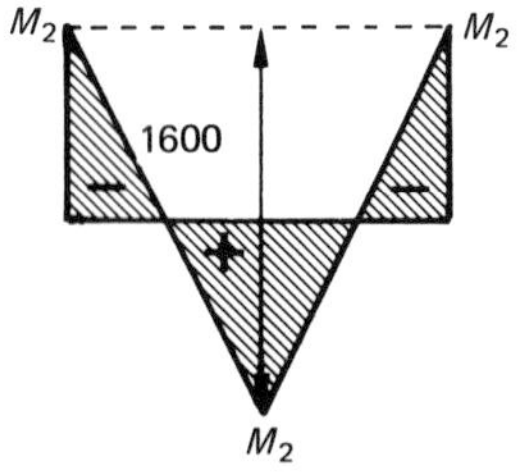

Figure 5.4

Span CD: guess that $M_3 < M_2$. Using similar triangles in the BMD, which is shown in figure 5.5

$$\frac{2}{3}M_3 + M_3 = 1813$$

$$M_3 = 1088 \text{ kN m} \tag{5.3}$$

Figure 5.5

The calculations are applications of the free and reactant BM method. The guesses were made to simplify the calculations, because as a result of the initial guess there was only one unknown in each span. Of course, the three guesses are incompatible; for example $M_1 < M_2$ in span AB, but $M_1 > M_2$ in span BC. The results of the calculations can now be used to check the guesses. Thus

$$M_2 \ (= 800 \text{ kN m}) < M_1 \ (= 1250 \text{ kN m})$$
$$< M_3 \ (= 1088 \text{ kN m})$$

so that the guess for span BC was correct, but those for AB and CD were not correct. The conclusion is that $M_2 = 800$ kN m but the calculations for AB and CD must be repeated using $M_B = M_C = 800$ kN m.

Span AB

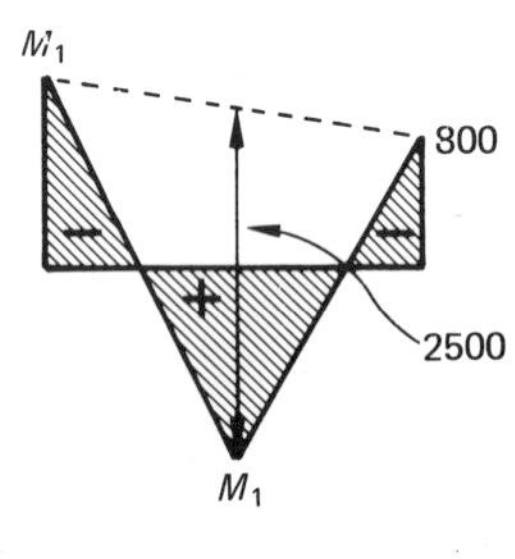

$$\frac{M_1 + 800}{2} + M_1 = 2500$$

$$\frac{3}{2}M_1 = 2100$$

$$M_1 = 1400 \text{ kN m} \tag{5.4}$$

Span CD

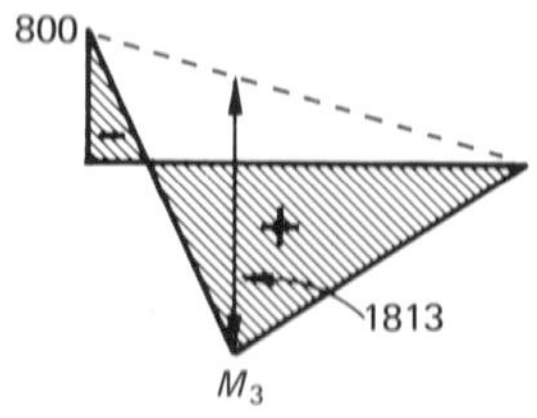

$$\frac{2}{3} \times 800 + M_3 = 1813$$

$$M_3 = 1280 \text{ kN m} \qquad (5.5)$$

Notice that the new calculations give *larger* values for M_1 and M_3, this is because effectively a smaller reactant moment has been used at B and C. This will always be the case, and is very reassuring because it means there is no possibility of the initially correct guesses becoming incorrect. Thus only one set of repeat calculations is ever necessary.

Plastic theory has given the required plastic moments in each span. Assuming mild steel (yield stress 250 N/mm^2) these can be transferred to member sizes using a steelwork properties handbook. [2]

Span	M_p (required) (kN m)	S (required) (cm^3)	Section chosen	S (provided) (cm^3)	Depth d (mm)
AB	1400	5600	686 x 254 UB 170	5616	692.9
BC	800	3200	533 x 210 UB 122	3198	544.6
CD	1280	5120	762 x 267 UB 147	5163	753.9

At this stage the designer can take charge and interpret this basic information as he wishes. This can be illustrated by considering various possibilities.

First Design

Accept the table of sections at face value and use a different section in each span. This design and its BMD at collapse are shown in figure 5.6. It is technically satisfactory, but a steel fabricator's nightmare. At B and C it is necessary to connect together members of completely different dimensions, in such a way that a plastic hinge will form in the weaker member. These connections are, of course, in addition to any others which may be necessary since it is unlikely that the steel would be available in 20 m lengths!

Second Design

To eliminate the connections at B and C a single size of member could be used throughout. This means that $M_1 = M_2 = M_3$. It is not necessary to do any more

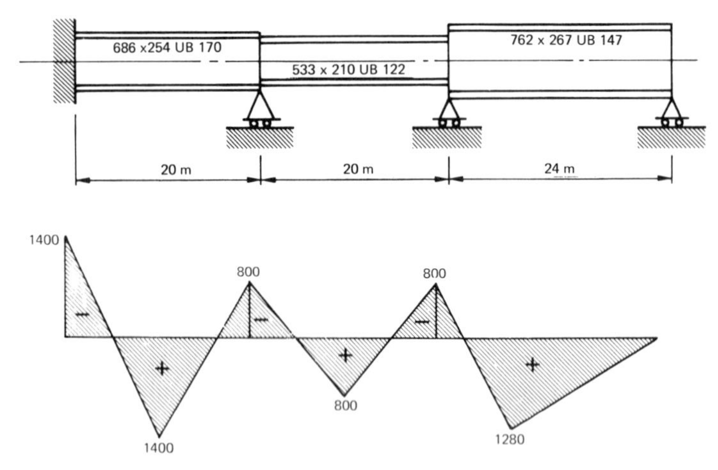

Figure 5.6

calculations because the initial results, equations 5.1 to 5.3, would be obtained. The largest value of plastic moment (M_1 = 1250 kN m) must then be used throughout. However, AB would collapse before the other spans since the plastic moment provided in them is considerably greater than required by the calculations. This is uneconomic use of material in BC and CD.

Third Design

Use the smallest possible member size (M_p = 800 kN m) in all spans. The BMD at collapse can be drawn by adding the same free BMs to the modified reactant BMs as in figure 5.7. Apart from two relatively short lengths in the middle of AB and CD the basic section provided has an adequate plastic moment. In those two short lengths the plastic moment can be increased by adding plates to the flanges of the basic section (see problem 2.1). The length of the plates is obtained from the BMD.

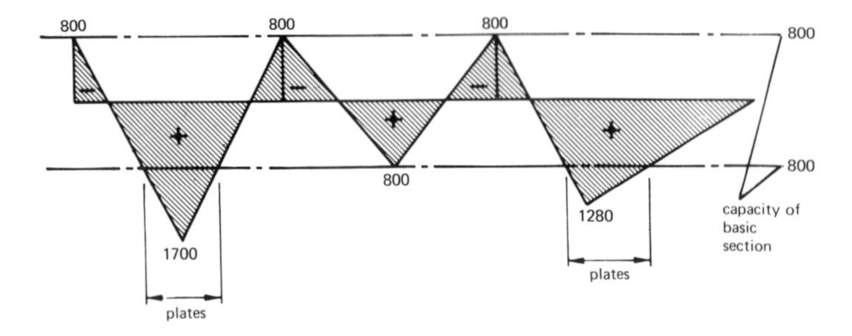

Figure 5.7

Plate details, span AB

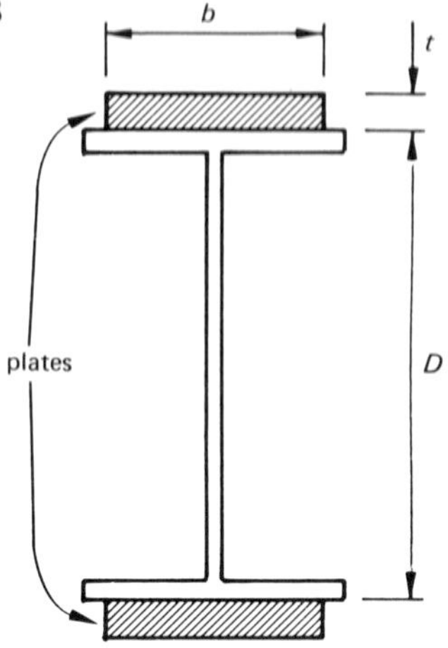

Figure 5.8

$$\text{maximum BM} = 1700 \text{ kN m}$$

$$\text{basic section} = \underline{800 \text{ kN m}}$$

$$M_p \text{ of plates} = 900 \text{ kN m}$$

$$M_p \text{ of plates} = btD\sigma_y$$

that is

$$bt = \frac{900 \times 10^6}{544.6 \times 250} = 6610 \text{ mm}^2$$

b and t can be chosen to suit. From similar triangles (figure 5.9)

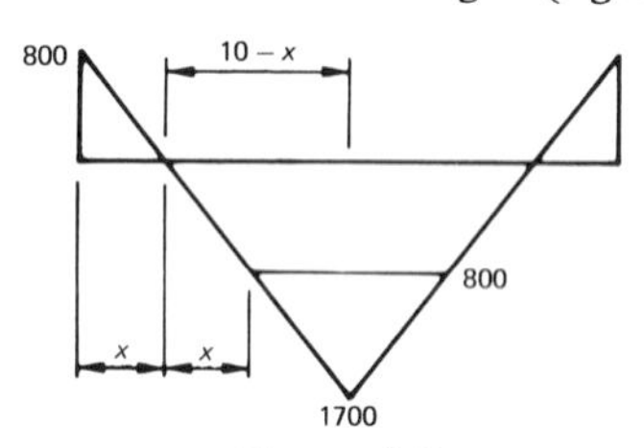

Figure 5.9

$$\frac{x}{10 - x} = \frac{800}{1700}$$

$$1700x = 8000 - 800x$$

$$x = \frac{8000}{2500} = 3.2 \text{ m}$$

$$\text{length of plates} = 2(10 - 2x)$$

$$= 2(10 - 6.4)$$

$$= 7.2 \text{ m}$$

Plate details, span CD

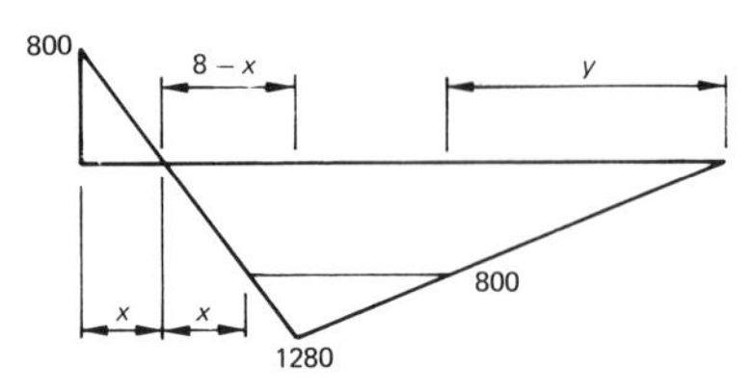

Figure 5.10

Maximum BM = 1280 kN m

basic section = 800 kN m

M_p of plates = 480 kN m

$$bt = \frac{480 \times 10^6}{544.6 \times 250} = 3526 \text{ mm}^2$$

From similar triangles (figure 5.10)

$$\frac{x}{8 - x} = \frac{800}{1280}$$

$$1280x = 6400 - 800x$$

$$x = \frac{6400}{2080} = 3.08 \text{ m}$$

$$\frac{y}{16} = \frac{800}{1280}$$

$$y = \frac{800 \times 16}{1280} = 10.0 \text{ m}$$

$$\text{length of plates} = (8 - 2x) + (16 - y)$$

$$= 24 - 2 \times 3.08 - 10$$

$$= 7.84 \text{ m}$$

Figure 5.11 shows the details of the design.

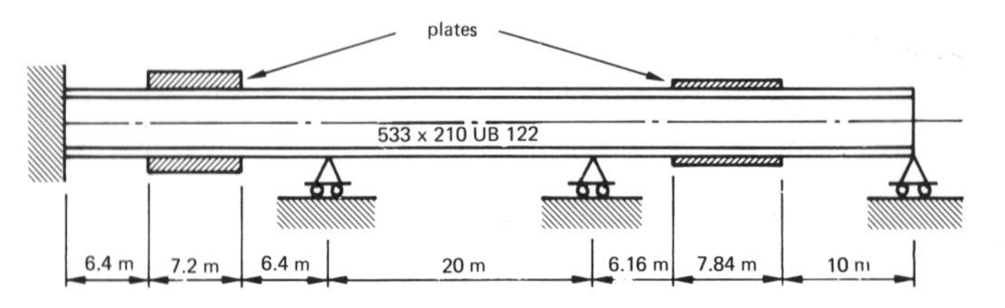

Figure 5.11

This example is not necessarily realistic in terms of loading, spans or supports. It is completely realistic in showing how plastic methods can be used to determine the overall size of the members and plates. It was possible to obtain the three designs with very little extra calculation. In a design by elastic methods it is usually necessary to make some assumptions about member sizes (usually second moments of area) in order to start the calculations, and only at the end can the assumptions be checked. With the plastic methods the initial guesses were very quickly checked and corrected. The evolution of the plated design was controlled by the designer. The plates have a significant effect on the elastic properties of the members, and this effect is very tedious to calculate. Figure 5.12 shows the elastic BMDs for the three designs under normal working load conditions. All three are satisfactory in that none of the BMs are large enough for yield to occur.

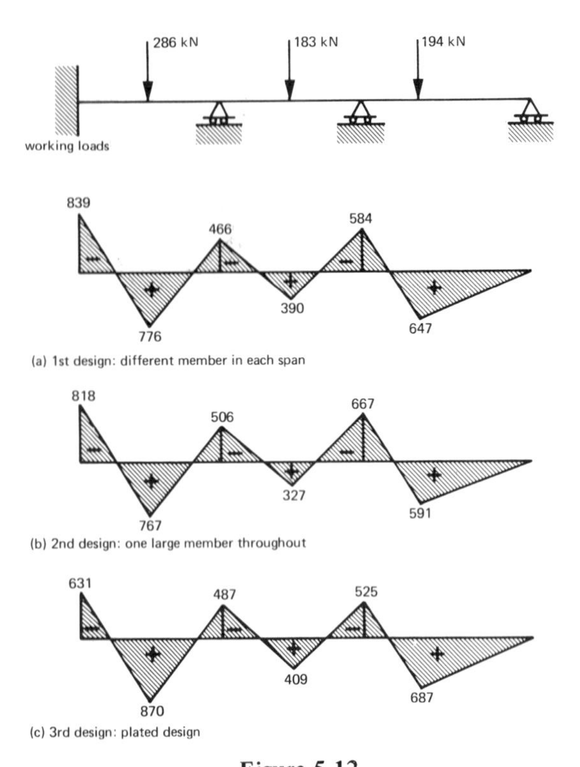

Figure 5.12

The dilemma is that there are three different designs which cannot be faulted technically. The choice between them is influenced by other factors. The first design can be rejected because it is needlessly difficult to fabricate. The choice between the other two will be decided by cost. Is the cost of fixing the plates to the beam (probably by welding) more or less than the cost of the extra steel required in the single member design?

The method outlined in the example was based on the free and reactant BMD method and is of course suitable for all beams. It must be emphasised that the method only gives the overall sizes of the members and plates. There are other details which would need to be considered, for example, web stiffeners at the supports and under the loads, but these are beyond the scope of this book.

5.4 OPTIMUM DESIGN

5.4.1 Factors Affecting a Design

The previous example showed how a continuous beam can be designed using plastic methods. Before moving on to consider framed structures it would be prudent to examine what the designer is trying to achieve.

The obvious objective is to produce a structure with a given load factor against collapse. As figure 5.13 shows, that in itself leads to a tightrope situation. If the collapse load factor of the final structure is greater than the required value, the structure is 'overdesigned' and therefore uneconomical. If the collapse load factor is smaller than the required value the structure is 'underdesigned' and unsafe. The designer has to get the design 'just right', achieving the minimum necessary strength.

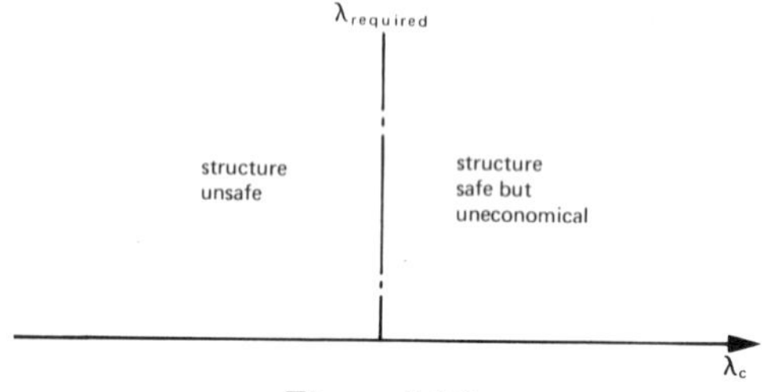

Figure 5.13

It usually turns out that there are several ways of doing this, and it is then up to the designer to choose the 'best' way. In other words he needs to achieve the *optimum design.*

To do this other factors than strength need to be examined. The most important factors are

(1) minimum total weight of material;

(2) availability of structural steel sections – there is no point in producing a design and then finding that the chosen members are unavailable;

(3) convenience of fabrication – machining, cutting and welding are all expensive;

(4) limiting deflections.

The optimum design will be a balance between these factors such that the final structure has

(5) minimum total cost.

It is now possible (thanks to the computer) to produce mathematically, within certain limitations, the optimum design. The full details are beyond the scope of this book, but it is possible to illustrate simply the approach. Most of the rest of the chapter will consider *minimum weight design*, designs which satisfy (1), but at the end it will be shown how the remaining factors can be considered.

Minimum weight designs by hand calculations require the use of interaction diagrams with the unknown plastic moments as the axes. This further limits the method to structures with only two different sized members. As will be seen there is virtually no limitation to the number of unknown members when the computer is used.

5.4.2 Weight Functions

The graph in figure 5.14 is a plot of weight per unit length against plastic modulus (M_p/σ_y) for all the universal beam and universal column sections detailed in the

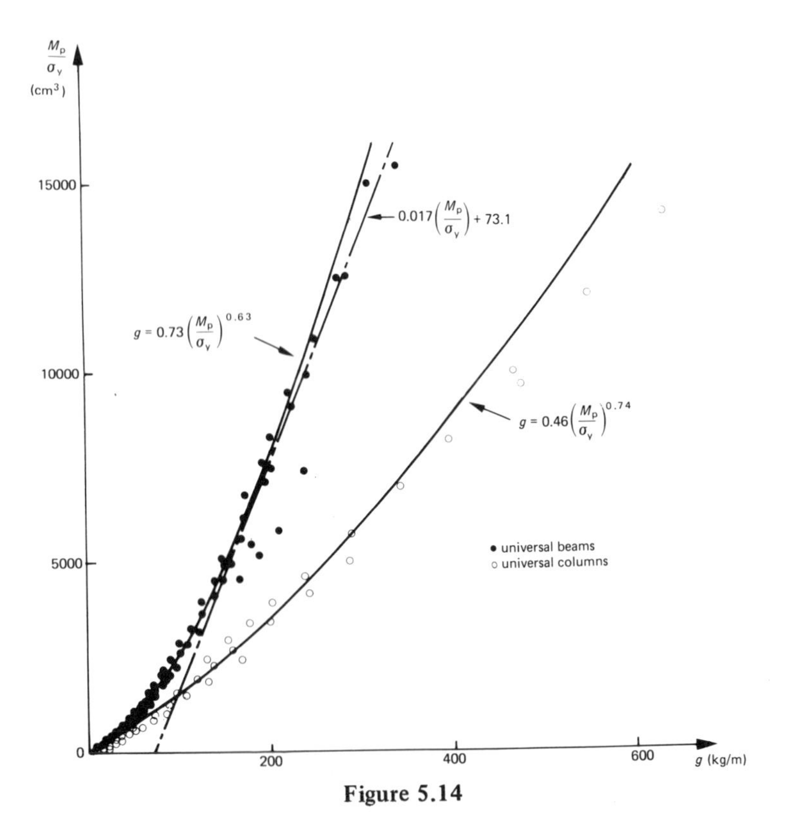

Figure 5.14

Structural Steelwork Handbook, [2] together with best fit curves through the points. There are distinct curves for the beams and the columns. In fact if this is done for all the possible section shapes there is a distinct curve for each shape. The curve is of the form

$$g = c\left(\frac{M_p}{\sigma_y}\right)^n \tag{5.6}$$

where g = weight/unit length, say kg/m, M_p = plastic moment, say kN m, and c, n are constants.

For sections made from the same material, σ has a constant value, so that

$$g = kM_p{}^n \tag{5.7}$$

where $k = c/(\sigma_y)^n$. This sort of relationship turns out to be difficult for hand calculations. In most designs it is unusual for the chosen members to be wildly different in size, most probably they would lie in the plastic moment range M_p to $2M_p$. Over such a range equation 5.7 can be replaced by a linear relationship of the form

$$g = k_1 M_p + k_2 \tag{5.8}$$

which is much more useful. Table 5.2 gives the k_1 and k_2 values for various S-ranges of the universal beam sections.

Table 5.2

S-range	Mild steel $\sigma_y = 250$ N/mm^2		Other steels	k_2
(cm^3)	M_p range (kN m)	k_1	k_1	
0 – 2000	0 – 500	0.154	$38.6/\sigma_y$	15.2
1000 – 3000	250 – 750	0.112	$28.1/\sigma_y$	30.7
2000 – 6000	500 – 1500	0.086	$21.6/\sigma_y$	48.0
5000 – 10 000	1250 – 2500	0.068	$17.0/\sigma_y$	73.1

The line for the S-range 5000 cm^3 to 10 000 cm^3 is shown in figure 5.14. The maximum discrepancy is when S is 5000 cm^3 and is only 1.4 per cent. Figure 5.14 shows that this line is in fact a very good representation of the *actual* relationship between g and S $(= M_p/\sigma_y)$ in the range 3000 cm^3 to 17 5000 cm^3.

The relationship between weight per unit length and plastic moment (equation 5.8 in general) can be used to find the total weight of the structure, G, in terms of the plastic moment of each member. This is called the *weight function* of the structure.

$$G = \Sigma L(k_1 M_p + k_2) \quad (5.9)$$

over
each
member

where L = member length

$$G = \Sigma k_1 M_p L + \Sigma k_2 L \quad (5.10)$$

In equation 5.10, the first part of the right-hand side $\Sigma k_1 M_p L$ is variable since its value depends on M_p, the second part is constant. The object of minimum weight design is *to make the variable part as small as possible.*

5.4.3 Minimum Weight Design of a Continuous Beam

A two-span continuous beam is the simplest example that can be used to illustrate minimum weight design. Figure 5.15 shows the beam. The first stage is to consider the possible collapse mechanisms. They must include the two possibilities $M_1 < M_2$ and $M_1 > M_2$ since the relative values of the moments are unknown.

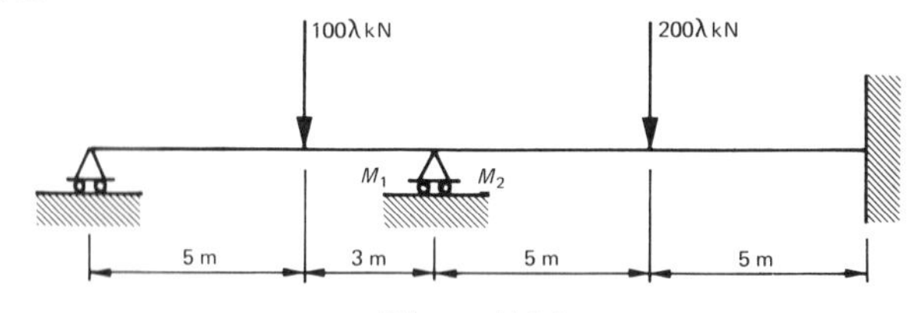

Figure 5.15

Mechanism	External Work	Internal Work $M_1 < M_2$	Internal Work $M_1 > M_2$
$3\phi = 5\theta$ $\phi = \frac{5}{3}\theta$	$100\lambda \times 5\theta$ $= 500\lambda\theta$	**a** $M_1\theta + 2M_1\phi$ $= M_1\left(1 + \frac{10}{3}\right)\theta$ $= \frac{13}{3}M_1\theta$	**b** $M_1\theta + M_1\phi + M_2\phi$ $= M_1\left(1 + \frac{5}{3}\right)\theta + M_2 \times \frac{5}{3}\theta$ $= \left(\frac{8}{3}M_1 + \frac{5}{3}M_2\right)\theta$
	$200\lambda \times 5\alpha$ $= 1000\lambda\alpha$	**c** $M_1\alpha + 3M_2\alpha$ $= (M_1 + 3M_2)\alpha$	**d** $4M_2\alpha$

The table shows the external and internal work for beam mechanisms in each span separately. There are four possible mechanisms, but only one of them is critical. Since the beam must be safe against all possible mechanisms, the work equation for each mechanism must be written in the form

$$
\begin{aligned}
&\frac{13}{3}M_1\theta \geqslant 500\lambda\theta && \mathbf{a} \\
&\left(\frac{8}{3}M_1 + \frac{5}{3}M_2\right)\theta \geqslant 500\lambda\theta && \mathbf{b} \\
&(M_1 + 3M_2)\alpha \geqslant 1000\lambda\alpha && \mathbf{c} \\
&4M_2\alpha \geqslant 1000\lambda\alpha && \mathbf{d}
\end{aligned}
\qquad (5.11)
$$

meaning that the internal resistance (represented by the internal work) is greater than or equal to effort of the applied loads, and so satisfying the safety requirement. In the (unknown) critical mechanism the work equation must be made an equality to make sure the structure has just the required load factor. The magnitude of the virtual rotations can be cancelled in each case. Equations 5.11 can be plotted on an interaction diagram with M_1 and M_2 as the axes. This is shown in figure 5.16, assuming that the required load factor is 2.0.

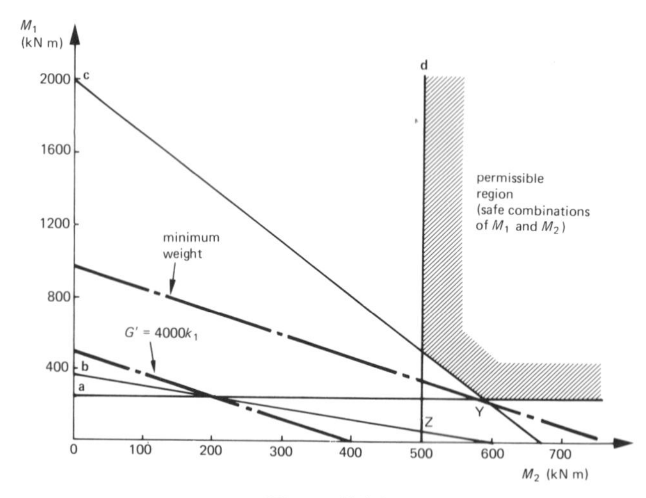

Figure 5.16

In contrast to chapter 3, the permissible region (PR) which satisfies equations 5.11 must be as shown in figure 5.16. The diagram shows that **b**, mechanism in the left-hand span with $M_1 > M_2$ is never critical, because it does not form a boundary of the PR. The intersection of **a** and **c**, point Y, represents the combination of M_1 and M_2 with $M_1 < M_2$ at which both spans would collapse simultaneously. There is no relationship between the plastic rotations in each span when this occurs, hence the reason for using different symbols in the original table. The intersection of **b** and **d** should represent simultaneous collapse with $M_1 > M_2$. In this case this is not possible because the intersection, point Z, shows $M_1 < M_2$.

Using equation 5.10 the weight function for the beam is

$$G = k_1(8M_1 + 10M_2) + 18k_2 \tag{5.12}$$

For the purposes of the design it is sufficient to consider the variable part of equation 5.12 only

$$G' = k_1 (8M_1 + 10M_2) \tag{5.13}$$

and add on $18k_2$ at the end.

Equation 5.13 is the equation of a family of parallel straight lines on the interaction diagram. One line can be drawn by guessing a value for G' and then finding two points on the line. Guess $G' = 4000k_1$, so that when

$$M_1 = 0 \qquad M_2 = 400 \text{ kN m}$$

$$M_2 = 0 \qquad M_1 = 500 \text{ kN m}$$

The line, which is shown in figure 5.16, lies outside the PR. If G' is increased sufficiently the line will just touch the boundary of the PR. This is the smallest value of G' which can be found from values of M_1 and M_2 which lie somewhere within the PR, and is called the *minimum weight line*. The values of M_1 and M_2 at the intersection of the minimum weight line and the boundary of the PR, point Y, are the member sizes for a minimum weight solution. The minimum weight design for the beam is

$$M_1 = 230.8 \text{ kN m}$$

$$M_2 = 589.7 \text{ kN m}$$

$$G'_{min} = k_1 (8 \times 230.8 + 10 \times 589.7) = 7743k_1$$

so that the actual minimum weight of material is

$$G_{min} = 7743k_1 + 18k_2$$

It turns out that the minimum weight design will result in simultaneous collapse in both spans, as is confirmed by the BMD at collapse shown in figure 5.17a. The minimum weight design requires a different size member in each span, which is perhaps not very practical. The alternative single section or plated designs can also be largely determined from the interaction diagram. The smallest possible value for M_2 is 500 kN m, determined by line **d**. If this section is used throughout the beam, the BMD at collapse is as shown in figure 5.17b. This shows that the section is satisfactory for both spans, and so determines the single section (esign. The smallest value of M_1 is 230.8 kN m from line **a**. The BMD at collapse, figure 5.1c, shows that plates with a plastic moment of 538.4 kN m would be required in the right-hand span.

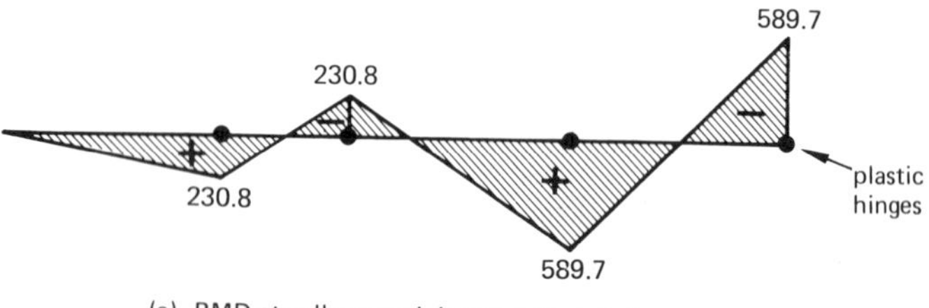

(a) BMD at collapse: minimum weight design

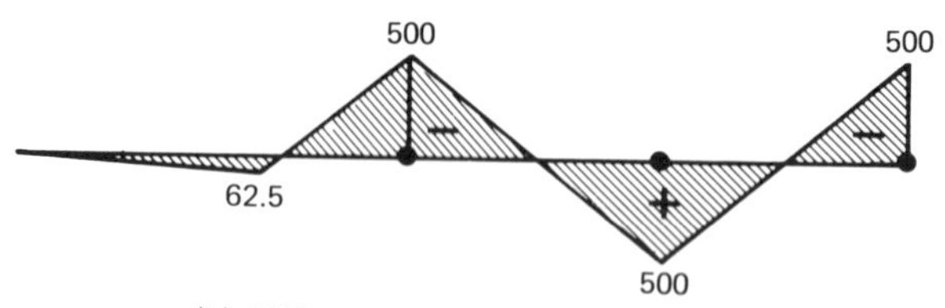

(b) BMD at collapse: single section design

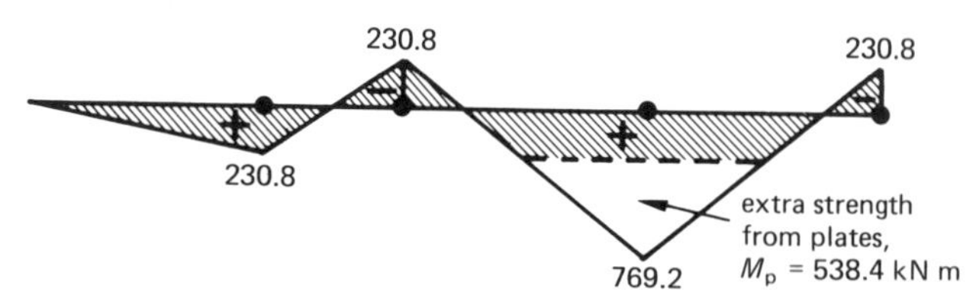

(c) BMD at collapse: plated design

Figure 5.17

5.4.4 Minimum Weight Design of a Portal Frame

Minimum weight design is more applicable to frames because it is relatively common and straightforward to connect different size members at right-angles (or near right-angles in pitched portal frames). The approach is identical to the continuous beam. The design of the portal frame in figure 5.18 illustrates how the method can be used when there are several load cases.

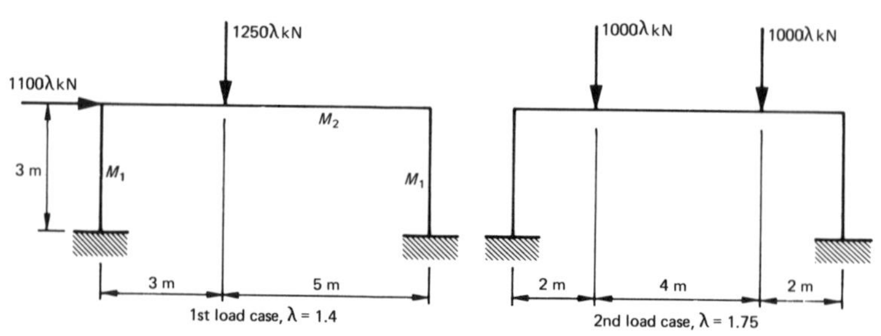

Figure 5.18

Mechanism	External Work	Internal Work $M_1 < M_2$	Internal Work $M_1 > M_2$
1st Load Case  $5\phi = 3\theta$ $\phi = \frac{3}{5}\theta$	$1250 \times 1.4 \times 3\theta$ $= 5250\theta$	**a** $M_1\theta + M_2(\theta + \phi) + M_1\phi$ $= (M_1 + M_2)(\theta + \frac{3}{5}\theta)$ $= \frac{8}{5}(M_1 + M_2)\theta$	**b** $2M_2(\theta + \phi)$ $= \frac{16}{5}M_2\theta$
	$1100 \times 1.4 \times 3\alpha$ $= 4620\alpha$	**c** $4M_1\alpha$	**d** $2(M_1 + M_2)\alpha$
 $\theta = \alpha$	$(5250 + 4620)\theta$ $= 9870\theta$	**e** $\frac{8}{5}(M_1 + M_2)\theta$ $+ 4M_1\theta - 2M_1\theta$ $= \left(\frac{18}{5}M_1 + \frac{8}{5}M_2\right)\theta$	**f** $\frac{16}{5}M_2\theta + 2(M_1 + M_2)\theta$ $- 2M_2\theta$ $= \left(2M_1 + \frac{16}{5}M_2\right)\theta$
2nd Load Case 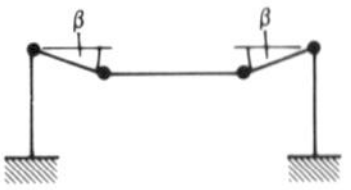 Note Symmetry	$2 \times 1000 \times 1.75 \times 2\beta$ $= 7000\beta$	**g** $2(M_1 + M_2)\beta$	**h** $4M_2\beta$
No horizontal load thus no sway			
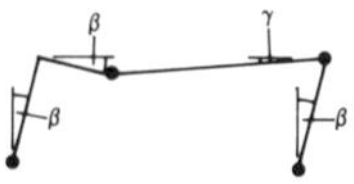 $6\gamma = 2\beta$ $\gamma = \frac{1}{3}\beta$	$1000 \times 1.75 \times (2\beta + 2\gamma)$ $= 3500\left(1 + \frac{1}{3}\right)\beta$ $= 4667\beta$	**j** $M_1(3\beta + \gamma) + M_2(\beta + \gamma)$ $= \left(\frac{10}{3}M_1 + \frac{4}{3}M_2\right)\beta$	**k** $M_1 \times 2\beta + M_2(2\beta + 2\gamma)$ $= \left(2M_1 + \frac{8}{3}M_2\right)\beta$

Note that the combined mechanism can occur without any horizontal force. The second hinge in the beam must disappear because the symmetry has been destroyed.

The inequalities resulting from the table are

$$\left.\begin{array}{ll} \frac{8}{5}M_1 + \frac{8}{5}M_2 \geqslant 5250 & \mathbf{a} \\ \frac{16}{5}M_2 \geqslant 5250 & \mathbf{b} \\ 4M_1 \geqslant 4620 & \mathbf{c} \\ 2M_1 + 2M_2 \geqslant 4620 & \mathbf{d} \\ \frac{18}{5}M_1 + \frac{8}{5}M_2 \geqslant 9870 & \mathbf{e} \\ 2M_1 + \frac{16}{5}M_2 \geqslant 9870 & \mathbf{f} \end{array}\right\} \text{1st Load Case} \qquad (5.14)$$

$$\left.\begin{array}{ll} 2M_1 + 2M_2 \geqslant 7000 & \mathbf{g} \\ 4M_2 \geqslant 7000 & \mathbf{h} \\ \frac{10}{3}M_1 + \frac{4}{3}M_2 \geqslant 4667 & \mathbf{j} \\ 2M_1 + \frac{8}{3}M_2 \geqslant 4667 & \mathbf{k} \end{array}\right\} \text{2nd Load Case} \qquad (5.15)$$

The interaction diagram from equations 5.14 and 5.15 is shown in figure 5.19. The PR satisfies the inequalities from both load cases.

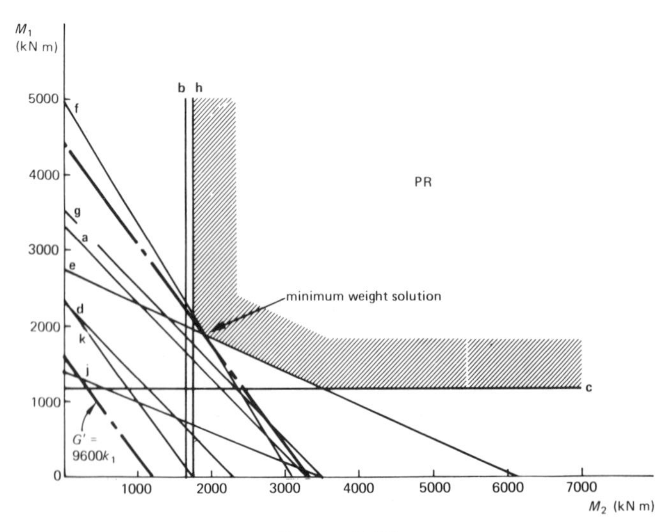

Figure 5.19

The weight function for the frame is

$$G = k_1(6M_1 + 8M_2) + 14k_2 \tag{5.16}$$

assuming the same type of members for columns and beam. The variable part of the weight function

$$G' = k_1 (6M_1 + 8M_2) \tag{5.17}$$

can be plotted on the interaction diagram to find the minimum weight solution. The solution is

$$M_1 = 1898 \text{ kN m}$$

$$M_2 = 1898 \text{ kN m}$$

$$G'_{min} = k_1(6 \times 1898 + 8 \times 1898) = 26572k_1$$

$$G_{min} = 26572k_1 + 13k_2$$

Notice that the minimum weight solution can be found without any knowledge of the constants k_1 and k_2 in the weight function. If the actual weight of material is required the relevant values of k_1 and k_2 can be found for the range of plastic moments in which M_1 and M_2 lie. These values are given in table 5.2 for the universal beam sections.

In conclusion, one note of warning. In the two examples a unique solution was found because the minimum weight line touches only one point on the boundary of the PR. Obviously in some cases they will be coincident and the solution is no longer unique. Any pair of plastic moments lying on the coincident boundary would be acceptable and give the same weight of material.

5.4.5 Minimum Weight Design by Computer

The object of minimum weight design is to minimise the structure weight function subject to the constraints of the inequalities derived from the possible mechanisms. The weight function and the constraints are all linear in the unknown plastic moments, so that the solution of the problem is one of *linear programming*. There are many schemes for achieving the solution, perhaps the most useful of which is the *Simplex* method. [9, 10] The Simplex method is reasonably efficient for use on a computer and can handle a large number of unknown plastic moments. There are standard computer programs in existence which will handle the complete process. [11, 12]

It has also proved possible to include some of the other design factors when using the computer. The starting point was to use the actual relationship between g and M_p, equation 5.7 when setting up the weight function. This made the function non-linear so that the solution had to be obtained using *non-linear programming*. [10] It then follows that the non-availability of sections can be considered by effectively removing the portion of equation 5.7 which corresponds to the unavailable sections. The weight function is then non-linear and non-continuous. [13]

The main difficulty with complicated frames is identifying the possible mechanisms. The design can only be correct if every possible constraint has been identified. In addition the number of constraints becomes very large so that the solution requires large amounts of computer time. Various formulations of the constraints have been used to reduce this. [10]

5.5 SUMMARY

The design of structures by plastic methods has been illustrated in this chapter. In continuous beams a straightforward trial and error process leads to designs which consider resistance to collapse and ease of fabrication. In frames, the purpose of the design process is to produce an optimum solution. In its simplest form this means a design requiring the use of the minimum weight of material to ensure the required load factor against collapse. A computer can be used for the design of more complex structures with many different member sizes and also to allow for other design considerations such as the non-availability of sections.

5.6 PROBLEMS

5.1 Determine, for the continuous beam in figure 5.20, the plastic moments for (a) a design based on the critical span, and (b) a plated design. The loads shown are the collapse loads.

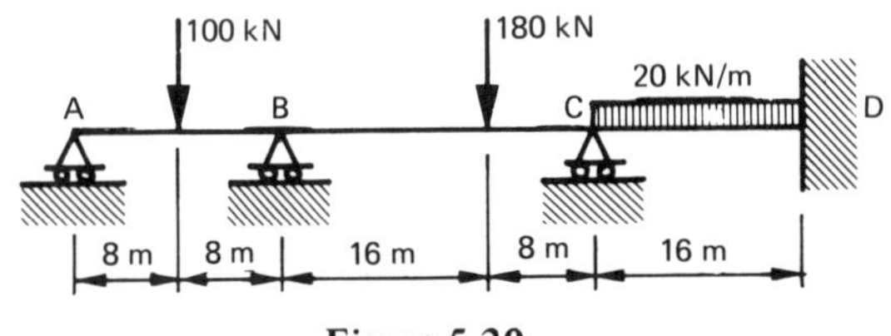

Figure 5.20

5.2 The continuous beam shown in figure 5.21 is to be designed with a load factor against collapse of 2.0. Examine the possible designs for (a) ease of fabrication (b) weight of material used. Which is the best design?

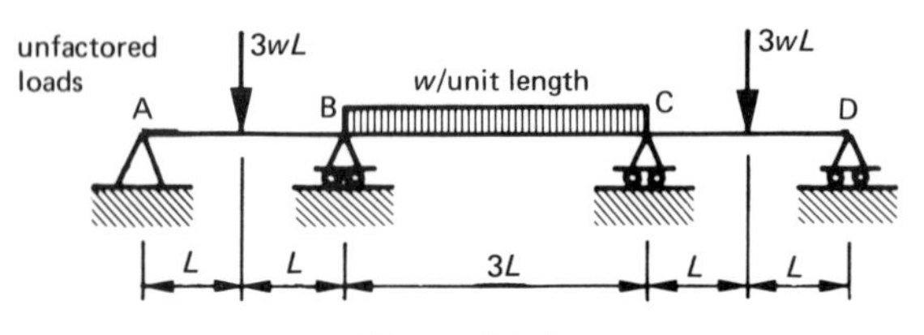

Figure 5.21

The weight per unit length of the beam section is kM_p, where M_p is the plastic moment of the section. The weight per unit length of plates is 1.5 kM, where M is the plastic moment of the plates. Where plates are used, they must extend $0.1L$ beyond the points where they are no longer required to resist bending.

5.3 Assuming the columns to be made from identical sections, what is the minimum weight design for the portal frame in figure 5.22? Assume the weight per unit length $g = k_1 M_p + k_2$.

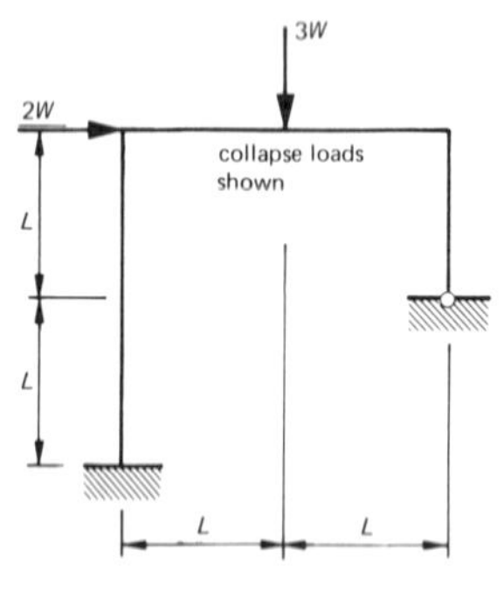

Figure 5.22

5.4 Design the pitched portal frame in figure 5.23 for minimum weight. Assume

for the columns $g_C = 0.75M_p + 90$

for the rafters $g_R = 0.4M_p + 75$

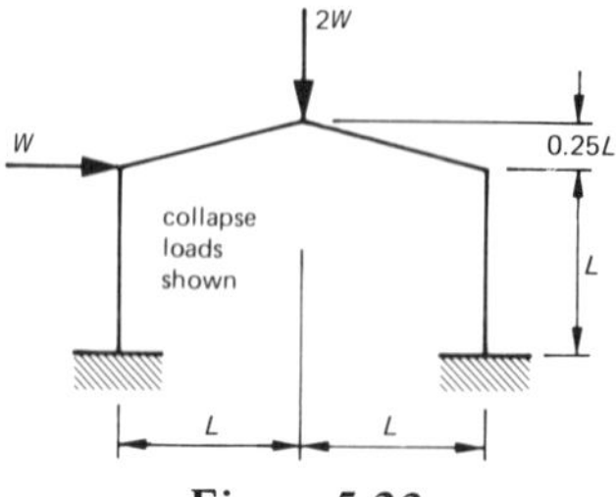

Figure 5.23

5.5 Assuming the same section for the columns, and another section for the sloping members, find the minimum weight of steel required in the frame shown in figure 5.24 if the weight per unit length is given by

$$g = 0.4M_p + 75 \text{ kg/m}$$

where M_p is in kN m.

The frame carries two load cases

(a) A horizontal wind force of 50 kN at the top of each column, and a vertical load of 200 kN at midspan. The required load factor is 1.5.

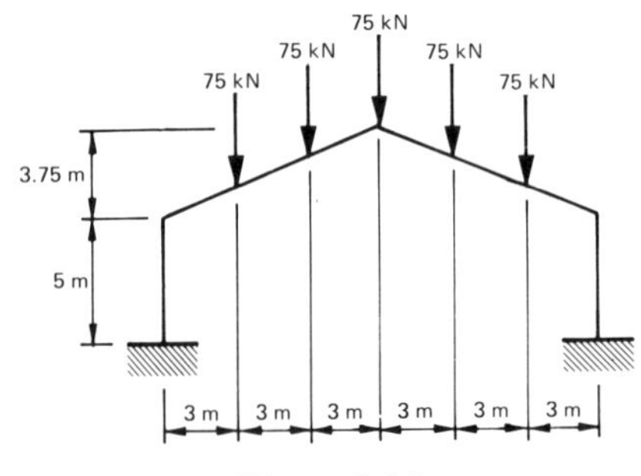

Figure 5.24

(b) A series of vertical loads only (as shown in the figure), with a load factor of 2.0.

5.6 Find the minimum weight design of the two-bay frame in figure 5.25, assuming one section for the columns and another for the beams. Use

$$g_C = 0.75M_p + 90 \text{ kg/m} \quad \text{for the columns}$$

$$g_B = 0.4M_p + 75 \text{ kg/m} \quad \text{for the beams}$$

where M_p is in kN m.

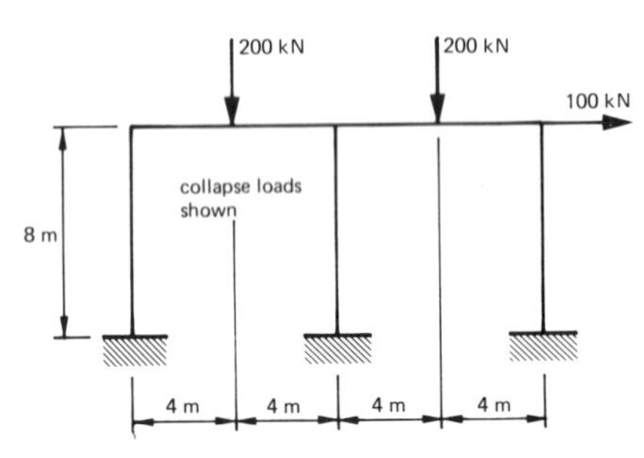

Figure 5.25

6 DEFLECTIONS AND STABILITY

6.1 INTRODUCTION

Methods for finding the collapse loads of steel frames were examined in detail in chapters 3, 4 and to some extent, 5. In the virtual work method, for example, the collapse load was found by considering small (virtual) deformations of the collapse mechanism. However, the shape of the structure before deformation of the mechanism was assumed to be the same as when there was no load on the structure. In other words, all deformation before collapse was ignored. There must be deformation before collapse, but how significant is it?

It is not sufficient to ensure that the structure is strong enough to resist the applied loading with an adequate load factor against collapse. It is also necessary to make certain that deflections do not become excessive. Consider the pitched portal frame shown in figure 6.1. The frame carries the rails for an overhead travelling crane. The crane-wheels which run on the rails will only have a finite amount of sideways movement. Consequently the dimension L is a critical part of the design. If it changes too much, due to deflection of the frame, the wheels will jam and the crane will no longer be travelling.

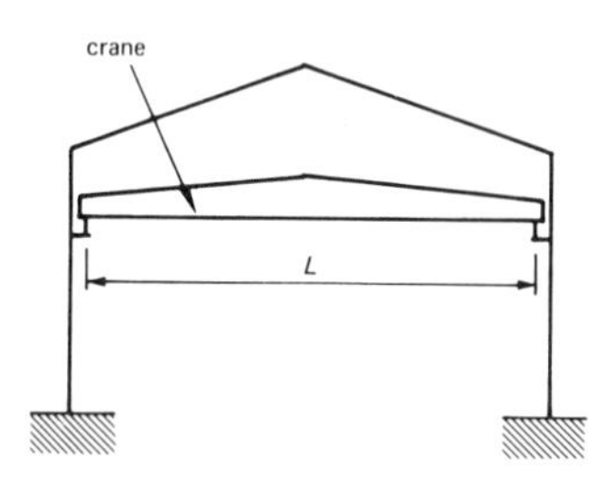

Figure 6.1

The deflections must be checked. Under normal working loads the structure should still be elastic so that it would be possible to find the deflections by elastic analysis. This is rather illogical in a structure designed by plastic methods: after all, one of the main reasons for using the plastic methods is that they avoid the tedious calculations of the elastic methods. There is another reason for being wary of the elastic methods. Figure 6.2 shows a possible load deflection curve for the frame in figure 6.1. Although the structure would be elastic at working loads, plastic design would produce a structure with bending moments close to the plastic moment at certain critical sections. It would only need a small overload on the crane for plastic hinges to form with a large increase in the deflections. Overhead cranes are notorious for overloading – after all, it is quicker to make one lift rather than two! A logical solution would be to limit the actual deflections at the point of collapse, especially if they can be calculated conveniently.

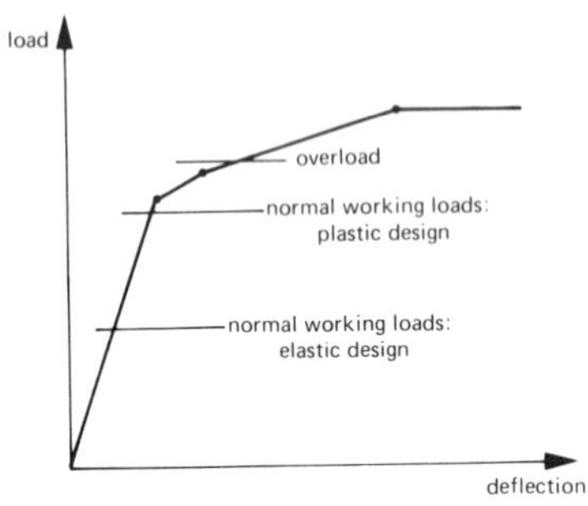

Figure 6.2

There is another possibility. The deflections before collapse may significantly reduce the collapse load of the structure. It is well known that deflections reduce the stiffness of struts (the so-called P–δ effect [14]) and the same thing can happen to frames. The problem is more serious in more flexible structures, where the deflections can cause an unexpected collapse mechanism due to overall buckling (instability) of the frame.

Originally only mild steel structures were designed by plastic methods, but nowadays higher strength steel structures are handled in the same way. The reduced ductility (see figure 1.5) is usually adequate for the formation of a collapse mechanism, but the higher yield stress means that smaller sections than would be required with mild steel can be used. This results in a more flexible structure with larger deflections. Obviously the problems caused by deflections are likely to be more severe. This must always be borne in mind when using higher strength steels.

The first part of this chapter describes a straightforward method for calculating deflections at the point of collapse. The second part is an examination of the effects of deflection on the collapse load.

6.2 CALCULATION OF DEFLECTIONS AT THE POINT OF COLLAPSE

6.2.1 Background Theory

One essential assumption that was used in finding collapse loads was that all plastic rotation occurs at the plastic hinges. This means that between the plastic hinges the members are elastic. Any frame can be broken down, therefore, into individual elastic members, with all plastic behaviour occurring at the ends of the members. (There will be some members, of course, whose end moments are less than the plastic moment.)

The deflections of the structure can be represented by the displacements of the ends of each member. Since the members are elastic, the end moments and the *elastic* end displacements can be related by the slope deflection equations.[1] Using the notation in figure 6.3 these are

$$M_{AB} = \frac{EI}{L}\left(4\theta_{AB} + 2\theta_{BA} - 6\frac{\delta}{L}\right) + (\mathrm{FEM})_{AB}$$
$$M_{BA} = \frac{EI}{L}\left(2\theta_{AB} + 4\theta_{BA} - 6\frac{\delta}{L}\right) + (\mathrm{FEM})_{BA} \qquad (6.1)$$

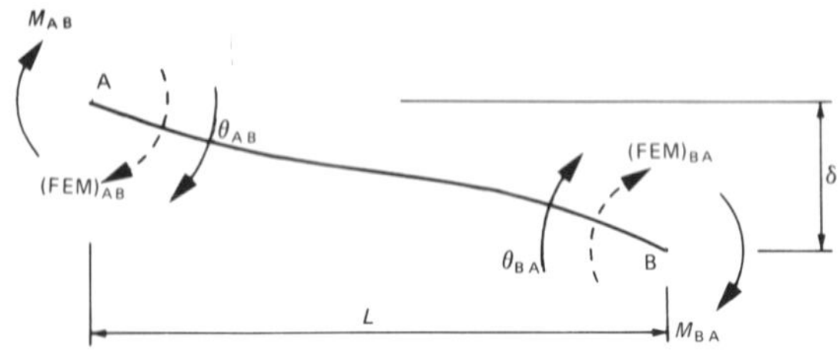

Figure 6.3

The diagram shows the positive sense of moments and deformations. This *clockwise positive convention* has been observed in the following examples. Equation 6.1 can be rearranged to give

$$\theta_{AB} = \frac{\delta}{L} + \frac{L}{6EI}(2M_{AB} - M_{BA}) - \frac{L}{6EI}[2(\mathrm{FEM})_{AB} - (\mathrm{FEM})_{BA}]$$
$$\theta_{BA} = \frac{\delta}{L} + \frac{L}{6EI}(-M_{AB} + 2M_{BA}) - \frac{L}{6EI}[-(\mathrm{FEM})_{AB} + 2(\mathrm{FEM})_{BA}] \qquad (6.2)$$

The modified slope deflection equations 6.2 can be used to find deflections at the point of collapse, when the final plastic hinge has just formed but has not started to rotate. It is easiest to explain the process by looking at examples.

6.2.2 Fixed End Beam with UDL

The beam and various stages of the analysis are shown in figure 6.4.

First Stage

The first stage of the analysis is to determine the collapse mechanism and collapse load (or load factor).

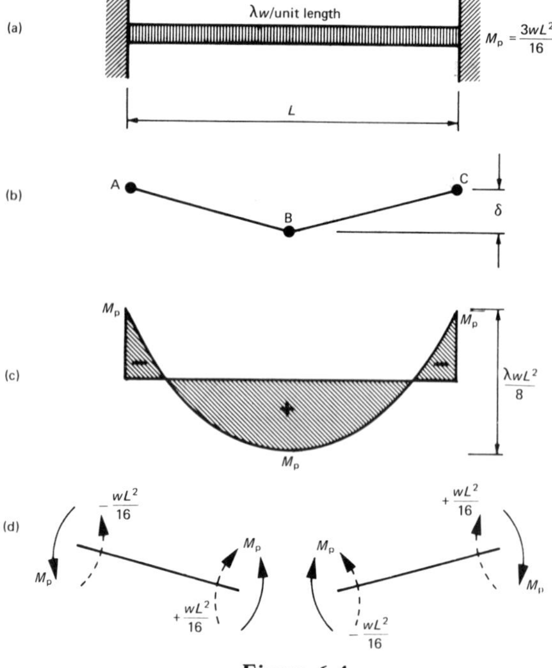

Figure 6.4

In this example, the problem is symmetric and the free and reactant BM method can be used. The mechanism and bending moment diagram are shown in figures 6.4b and c. From the geometry of the BMD

$$\frac{\lambda_c wL^2}{8} = 2M_p$$

so that when $M_p = 3wL^2/16$

$$\lambda_c = 3$$

Second Stage

It is now necessary to divide the structure into individual (elastic) members and write down the modified slope deflection equations for each member.

The beam can be divided into two members, AB and BC, between the plastic hinges. The end moments for both members are equal to the plastic moment M_p. The difficulty is to determine the direction in which they act. This can be done by noting that the end moments resist the end rotations, therefore they must act in *the opposite sense to the plastic rotations.* The fixed end moments in AB and BC are the standard case of a UDL ($3w$ per unit length) on a fixed beam (of span $L/2$). The various moments are summarised in figure 6.4d.

The slope deflection equations can now be written down for each member

$$\theta_{AB} = \frac{2\delta}{L} + \frac{L}{12EI}(-2M_p + M_p) - \frac{L}{12EI}\left(-\frac{wL^2}{8} - \frac{wL^2}{16}\right)$$

$$\theta_{AB} = \frac{2\delta}{L} - \frac{M_pL}{12EI} + \frac{wL^3}{64EI}$$

$$\theta_{BA} = \frac{2\delta}{L} + \frac{L}{12EI}(M_p - 2M_p) - \frac{L}{12EI}\left(\frac{wL^2}{16} + \frac{wL^2}{8}\right)$$

$$\theta_{BA} = \frac{2\delta}{L} - \frac{M_pL}{12EI} - \frac{wL^3}{64EI}$$

$$\theta_{BC} = -\frac{2\delta}{L} + \frac{L}{12EI}(2M_p - M_p) - \frac{L}{12EI}\left(-\frac{wL^2}{8} - \frac{wL^2}{16}\right)$$

$$\theta_{BC} = -\frac{2\delta}{L} + \frac{M_pL}{12EI} + \frac{wL^3}{64EI}$$

and similarly

$$\theta_{CB} = -\frac{2\delta}{L} + \frac{M_pL}{12EI} + \frac{wL^3}{64EI}$$

Notice the term $-2\delta/L$ in the last two equations. The sign convention in figure 6.3 defined the deflection δ as positive when the right-hand end (B) sank below the left-hand end (A), causing a clockwise rotation of the whole member. In the example, BC is rotating anticlockwise, hence the negative sign.

Third Stage

The deflection must now be calculated when the last plastic hinge has just formed. But which is the last hinge? There is no way of knowing this, so each hinge in turn must be assumed to form last, and a deflection calculated for each one.

If the hinge at A (or at C because of symmetry) is the last to form, there will have been no rotation at A (or C) because it is a clamped end. Thus

$$\theta_{AB} = \frac{2\delta}{L} - \frac{M_pL}{12EI} + \frac{wL^3}{64EI} = 0$$

so that

$$\frac{2\delta}{L} = \frac{M_pL}{12EI} - \frac{wL^3}{64EI} = \frac{wL^3}{EI}\left(\frac{3}{12.16} - \frac{1}{64}\right)$$

$$\delta = 0$$

If the hinge at B is the last to form, there will have been no plastic rotation at B at the point of collapse. The whole beam (AC) will still be continuous at B. This can only be achieved if

$$\theta_{BA} = \theta_{BC}$$

Substituting for θ_{BA} and θ_{BC} gives

$$\frac{2\delta}{L} - \frac{M_pL}{12EI} - \frac{wL^3}{64EI} = -\frac{2\delta}{12EI} + \frac{wL^3}{64EI}$$

$$\frac{4\delta}{L} = \frac{2M_pL}{12EI} + \frac{2wL^3}{64EI} = \frac{2wL^3}{EI}\left(\frac{3}{12.16} + \frac{1}{64}\right)$$

$$= \frac{4wL^3}{64EI}$$

$$\delta = \frac{wL^4}{64EI}$$

The question now is which of these values is correct?

Fourth Stage

One way to decide is to substitute the values of δ back into the slope deflection equations and obtain the rotations.

The results of this are summarised in table 6.1. Inspection of the table shows that the first deflected shape is ridiculous. The only conclusion is that the last hinge forms at B and the deflection at the point of collapse is $wL^4/64EI$. In a more complicated structure it would be tedious to set up a much enlarged version of table 6.1. Instead the deflection can be chosen by using the displacement theorem. This states *'Let displacements be predicted on the basis of each plastic hinge forming last. If in the loading process* no *hinge once formed has been unloaded, the largest displacement so predicted will be the correct one.'*

Table 6.1

Last hinge	δ	θ_{AB}	θ_{BA}	θ_{BC}	θ_{CB}	Deflected shape
A or C	0	0	$-\frac{wL^3}{32EI}$	$+\frac{wL^3}{32EI}$	0	plastic hinge
B	$\frac{wL^4}{64EI}$	$+\frac{wL^3}{32EI}$	0	0	$-\frac{wL^3}{32EI}$	

Since the calculations are based on conditions at collapse there can be no indication of any hinge which may have formed and then disappeared. However, it is unusual for that to happen. Thus the largest deflection is normally correct.

6.2.3 Portal Frame Example

This second example brings out two more important points. The analysis is summarised in figure 6.5.

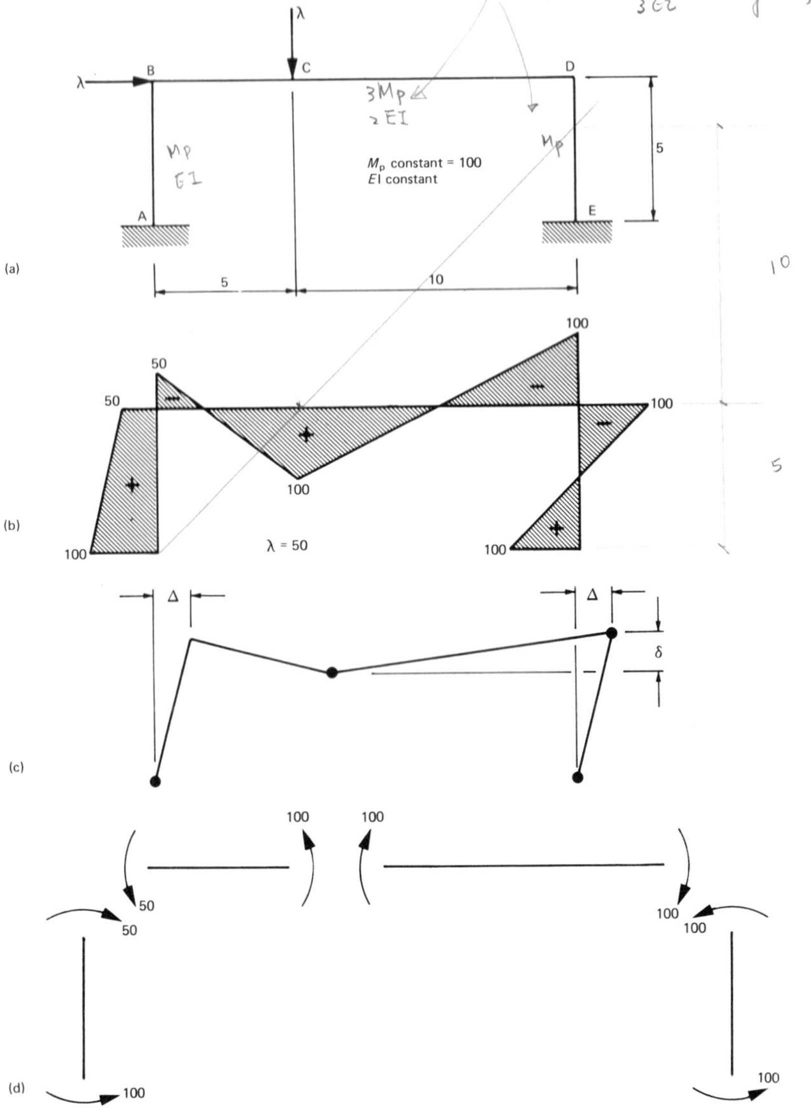

Figure 6.5

First Stage

The virtual work method shows that the frame collapses by the combined mechanism when $\lambda_c = 50$. The BMD at collapse and the collapse mechanism are shown in figures 6.5b and c.

Second Stage

The structure can be broken down into four members as in figure 6.5d. There are no fixed end moments in this case because the point loads are at the ends of members. There is no problem in deciding the direction of the end moments except at B. Here there is no plastic hinge in the mechanism. To decide on the direction of the end moments, imagine that the BM at B is increased until a hinge forms. The end moments act to resist the rotation of that imaginary hinge, as shown in figure 6.6.

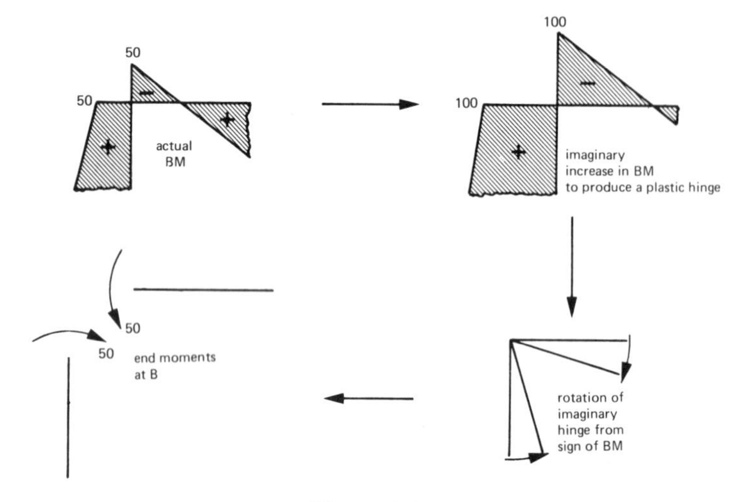

Figure 6.6

Two deflections are required in order to write down the slope deflection equations. It is assumed that the vertical deflection in the beam, δ, is small so that the tops of both columns move Δ horizontally. (This is the same assumption as used in calculating collapse loads.) The slope deflection equations are

$$\theta_{AB} = \frac{\Delta}{5} + \frac{5}{6EI}(-200 - 50) = \frac{\Delta}{5} - \frac{1250}{6EI}$$

$$\theta_{BA} = \frac{\Delta}{5} + \frac{5}{6EI}(+100 + 100) = \frac{\Delta}{5} + \frac{1000}{6EI}$$

$$\theta_{BC} = \frac{\delta}{5} + \frac{5}{6EI}(-100 + 100) = \frac{\delta}{5}$$

$$\theta_{CB} = \frac{\delta}{5} + \frac{5}{6EI}(50 - 200) = \frac{\delta}{5} - \frac{750}{6EI}$$

$$\theta_{CD} = -\frac{\delta}{10} + \frac{10}{6EI}(200 - 100) = -\frac{\delta}{10} + \frac{1000}{6EI}$$

$$\theta_{DC} = -\frac{\delta}{10} + \frac{10}{6EI}(-100 + 200) = -\frac{\delta}{10} + \frac{1000}{6EI}$$

$$\theta_{DE} = \frac{\Delta}{5} + \frac{5}{6EI}(-200 + 100) = \frac{\Delta}{5} - \frac{500}{6EI}$$

$$\theta_{ED} = \frac{\Delta}{5} + \frac{5}{6EI}(100 - 200) = \frac{\Delta}{5} - \frac{500}{6EI}$$

Third Stage

At point B the structure is still elastic. In order to maintain continuity at B

$$\theta_{BA} = \theta_{BC}$$

Substituting for θ_{BA} and θ_{BC} gives

$$\frac{\Delta}{5} + \frac{1000}{6EI} = \frac{\delta}{5}$$

which is a relationship between the two unknown deflections.
Each hinge in turn must now be assumed to form last.
A forms last. $\theta_{AB} = 0$ fixed support at A, that is

$$\Delta = \frac{1041.7}{EI} \qquad \delta = \frac{1875}{EI}$$

C forms last. $\theta_{CB} = \theta_{CD}$ to maintain continuity at C, that is

$$\frac{\delta}{5} - \frac{750}{6EI} = -\frac{\delta}{10} + \frac{1000}{6EI}$$

$$\delta = \frac{972.2}{EI} \qquad \Delta = \frac{138.9}{EI}$$

D forms last. $\theta_{DC} = \theta_{DE}$ to maintain continuity at D

$$-\frac{\delta}{10} + \frac{1000}{6EI} = \frac{\Delta}{5} - \frac{500}{6EI}$$

substitute for δ to give

$$-\frac{\Delta}{10} - \frac{500}{6EI} + \frac{1000}{6EI} = \frac{\Delta}{5} - \frac{500}{6EI}$$

$$\Delta = \frac{555.6}{EI} \qquad \delta = \frac{1388.9}{EI}$$

E forms last. $\theta_{ED} = 0$ fixed support at E

$$\Delta = \frac{416.7}{EI} \qquad \delta = \frac{1250}{EI}$$

Fourth Stage

The largest displacements occur when the hinge at A forms last. Hence at the point of collapse

$$\Delta = \frac{1041.7}{EI} \qquad \delta = \frac{1875}{EI}$$

This frame was used in chapter 3 to illustrate the gradual formation of plastic hinges. The results of that analysis were obtained by a step-by-step stiffness analysis using a computer. The results of that analysis (figure 3.2) are identical to the ones obtained here.

6.2.4 Sloping Members

Structures with sloping members need careful attention. Figure 6.7 shows a typical pitched portal frame. The problem is that, as with the pitched portal

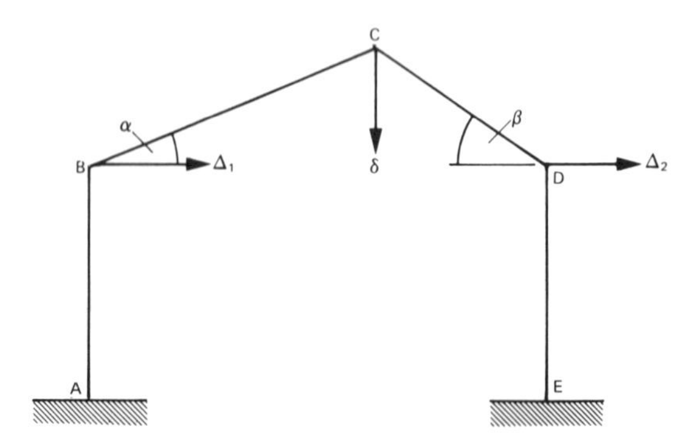

Figure 6.7

mechanism, the tops of the columns do not deflect the same amount. An extra equation relating these deflections must be found. From the figure

$$\Delta_2 = \Delta_1 + \delta(\tan \alpha + \tan \beta) \tag{6.3}$$

where δ is the vertical deflection at C. Remember that in the slope deflection equations the deflection normal to the sloping members (that is, $\delta \sec \alpha$ and $\delta \sec \beta$) must be used.

6.3 THE EFFECT OF DEFLECTION ON THE COLLAPSE LOAD

The examples in the previous section showed that there can be significant deflections before collapse starts. Deflections, particularly in columns with substantial compressive axial forces, can cause serious instability (buckling) in frames. In this section the effect of this on the collapse load is examined by means of two examples. A practical method of allowing for instability is then outlined.

6.3.1 Horne's Example of a Cantilever Column

Horne [3] has given an excellent illustration of the effect that deflections can have on the collapse load. His example is repeated here in a slightly extended form.

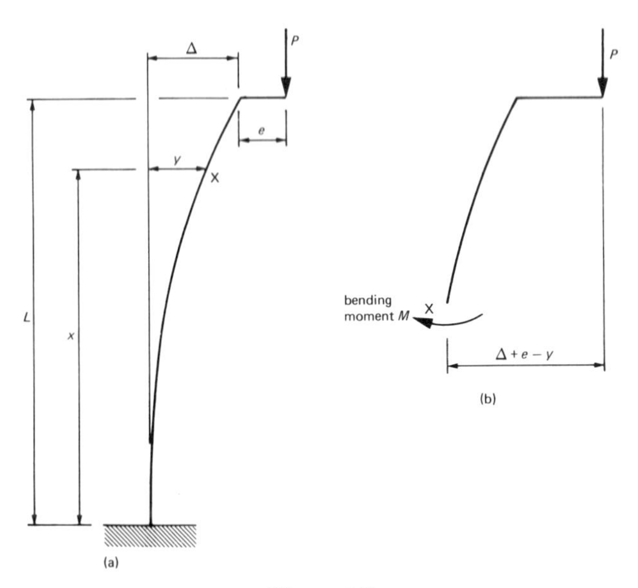

Figure 6.8

Figure 6.8a shows a cantilever column, rigidly clamped at the base and free at the top. An eccentric vertical force is applied to the column, causing the column to bend and deflect sideways. Column properties are

length $L = 2$ m

eccentricity $e = 0.1$ m

Young's modulus $E = 200$ kN/mm^2

yield stress $\sigma_y = 250$ N/mm^2

Cross-section is square, side length $d = 0.1$ m thus

$$I = \frac{d^4}{12} = 8.333 \times 10^{-6}\ \text{m}^4$$

$$Z = \frac{d^3}{6} = 1.667 \times 10^{-4}\ \text{m}^3$$

$$S\ (\text{plastic modulus}) = \frac{d^3}{4} = 2.5 \times 10^{-4}\ \text{m}^3$$

$$A\ (\text{cross-sectional area}) = \text{d}^2 = 0.01\ \text{m}^2$$

Consider first the elastic behaviour of the column. Figure 6.8b shows the free body diagram from cutting the column at some point X. Moment equilibrium about X gives

$$M = -P(\Delta + e - y)$$

Using the moment curvature relationship of bending theory

$$EI\frac{\text{d}^2y}{\text{d}x^2} = -M = P(\Delta + e - y)$$

which can be rearranged in the form

$$\frac{\text{d}^2y}{\text{d}x^2} + \alpha^2 y = \alpha^2(\Delta + e) \tag{6.4}$$

where $\alpha^2 = P/EI$. This differential equation governs the deflections (y) of the column. The solution of this equation has been shown [15] to be

$$y = (\Delta + e)(1 - \cos \alpha x) \tag{6.5}$$

Substituting $x = L$ and $y = \Delta$ into equation 6.5 gives the deflection Δ at the top of the column

$$\Delta = e(\sec \alpha L - 1)$$

Replacing α and substituting the column properties gives

$$\Delta = 0.1\ [\sec (1.55 \times 10^{-3}\sqrt{P}) - 1] \tag{6.6}$$

The load deflection relationship ($P - \Delta$) defined by equation 6.6 is shown in figure 6.9. Although the analysis assumed elastic behaviour the curve is non-linear because of increasing instability in the column. Apparently Δ becomes infinitely large when

$$\sec (1.55 \times 10^{-3}\sqrt{P_\text{E}}) = \infty$$

$$1.55 \times 10^{-3}\sqrt{P_\text{E}} = \frac{\pi}{2}$$

$$P_\text{E} = 1.027 \times 10^6\ \text{N}$$

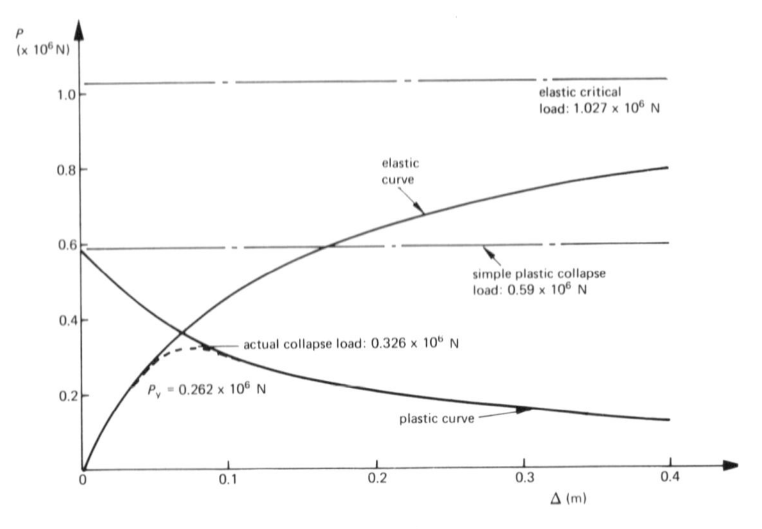

Figure 6.9

It can be shown that when the eccentricity, e, is zero, the strut would remain straight until this load, when it would buckle sideways. P_e is called the *buckling* or *elastic critical* load of the column.

The stress in the column is a combination of axial and bending stresses. The largest stress is at the base, where the BM is greatest. When this stress reaches the yield stress the elastic analysis will cease to be correct. This will be when

$$\frac{P}{A} + \frac{P(\Delta + e)}{Z} = \sigma_y$$

substituting for A, Z, Δ, e and σ_y gives

$$100P + 600P \sec (1.55 \times 10^{-3}\sqrt{P}) = 250 \times 10^6$$

This rather complicated equation can be solved by trial and error to show that the load at first yield is

$$P_y = 0.262 \times 10^6 \text{ N}$$

It is possible, but complicated, to trace the spread of yield, but it is more convenient to look now at the collapse of the column. Figure 6.10 shows the collapse mechanism. The column becomes a lever rotating about a plastic hinge at the base. The reduced plastic moment (allowing for the axial force) of the column resists the rotation. From section 2.5

$$\frac{M_p'}{M_p} = 1 - \left(\frac{P_c}{P_p}\right)^2$$

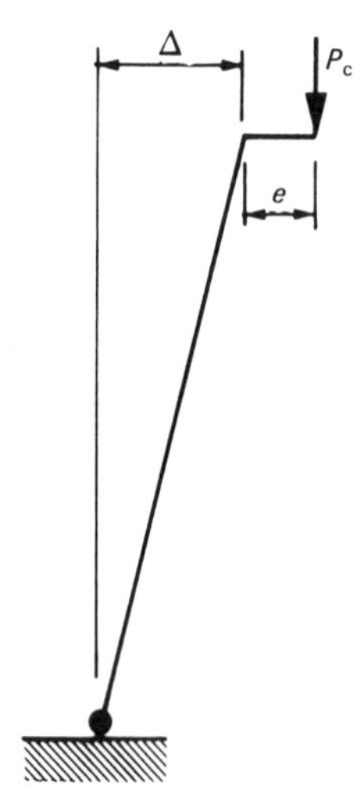

Figure 6.10

for a rectangular section. Using

$$M_p = S\sigma_y = \frac{d^3\sigma_y}{4}$$

and

$$P_p = A\sigma_y = d^2\sigma_y$$

and putting $P_c = nP_p$ gives

$$M_p' = (1 - n^2) \cdot \frac{d^3\sigma_y}{4}$$

The moment of the collapse load P_c about the base causes the rotation.

disturbing moment = $(\Delta + e) P_c = (\Delta + e) nd^2 \sigma_y$

At the point of collapse the disturbing and resisting moments are equal, because the column is in equilibrium. At collapse then

$$(\Delta + e)nd^2\sigma_y = (1 - n^2)\frac{d^3\sigma_y}{4}$$

which can be rearranged into

$$n^2 + \frac{4}{d}(\Delta + e)n - 1 = 0$$

The solution of this quadratic equation gives the collapse load P_c as a proportion of P_p, but the solution depends on the deflection Δ at the top of the column. The collapse loads at various values of Δ have been plotted in figure 6.9.

Figure 6.9 summarises the behaviour of the column from zero load until collapse. The broken line shows approximately the transition from elastic behaviour as yield spreads through the base of the column. There are two points to note.

(1) There is substantial deflection before collapse occurs.

(2) A simple plastic collapse calculation would have used the mechanism in figure 6.10 but with $\Delta = 0$. That load (0.590×10^6 N) is considerably greater than the true collapse load (about 0.325×10^6 N).

This is a rather extreme example, but it does illustrate the effects of deflection before collapse. In any frame where there are compressive axial forces in the columns, the true collapse load is smaller than the collapse load predicted by simple plastic analysis. The reduction is usually less marked than shown here (as can be seen in the next example) but cannot be ignored.

6.3.2 Portal Frame Example

A similar analysis for a portal frame is more complicated. The approach is similar to that described in section 3.2 using a stiffness analysis and modifying the structure each time a plastic hinge forms. However, the stiffness matrices for the columns must be formed using the stability functions [14] to allow for the effect of axial load on stiffness. The analysis at each load factor now requires an iterative procedure to arrive at the solution, and is both complicated and expensive in computer time.

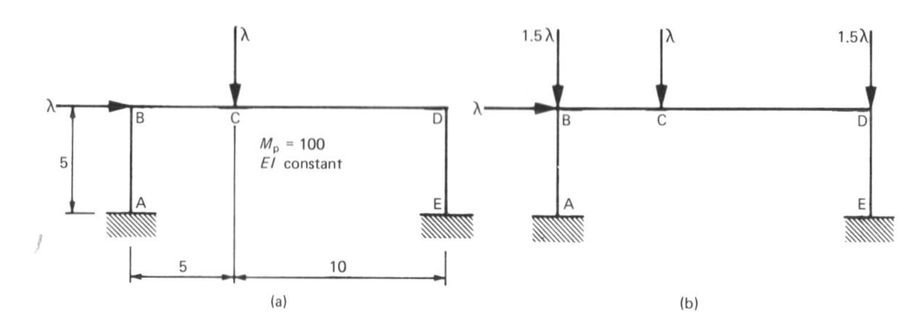

Figure 6.11

The frame shown in figure 6.11a was analysed in this way and the results are given in figure 6.12. The frame is identical to the one used in sections 3.2 and 6.2.3. In the original analysis (section 3.2), the effects of axial load and deflection were ignored, and the collapse load factor was 50.0. The frame collapsed by the combined mechanism with the hinges forming in the order E, C, D and A. The three load deflection curves in figure 6.12 were obtained by using different *EI*

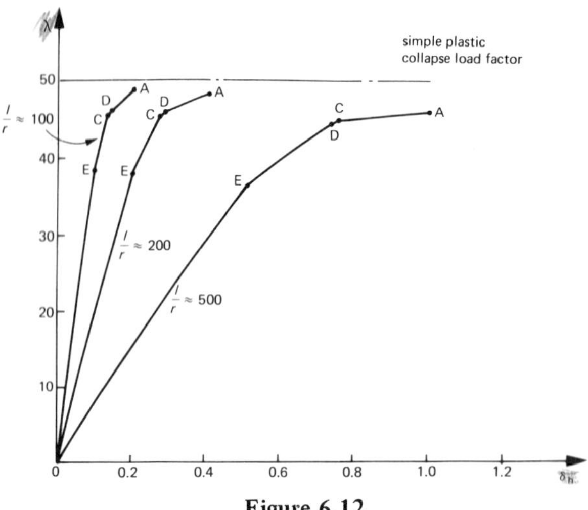

Figure 6.12

values for the members. These were chosen to give slenderness ratios, l/r, for the columns of approximately 100, 200 and 500. The axial loads cause non-linear behaviour between the formation of each plastic hinge, but more importantly, they also reduce the collapse load factor. The bigger the deflections in the structure the bigger the reduction. In each case the final mechanism is the combined one, but in the very flexible frame ($l/r = 500$) the deflections modify the order in which the hinges form. Table 6.2 shows the collapse load factors. The biggest reduction is about 10 per cent although the corresponding l/r ratio of 500 is much greater than would be used in a practical frame. The l/r ratio of 200 is about the practical limit, and in that case, the reduction is about 4 per cent.

Table 6.2

Slenderness ratio (l/r)	100	200	500
Reduced collapse load factor	49.10	48.25	45.93

The reduction in collapse load factor is not really too much of a problem in single-storey frames, but can be much more so in multi-storey frames. Multi-story behaviour has been simulated by applying extra loads to the columns of the portal frame as shown in figure 6.11b. The extra loads represent the weight of the structure and the loading in the higher storeys. As can be seen in figure 6.13 the results are more dramatic. The higher axial loads cause bigger deflections and significant changes in behaviour. Table 6.3 summarises the collapse loads and

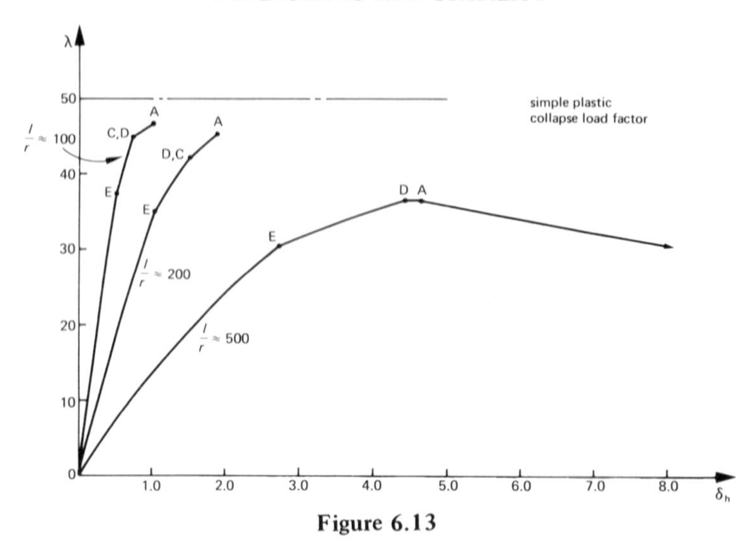

Figure 6.13

mechanisms. (Actually the analysis results are open to question because the deflections are no longer 'small', but the results do give at least a qualitative picture of what can happen.)

Table 6.3

Slenderness ratio	Collapse load factor	Collapse mechanism
100	46.67	Plastic collapse, combined mechanism
200	45.13	Plastic collapse combined mechanism
500	36.60	Sway buckling

In the stiffer frames ($l/r \leqslant 200$) collapse still occurs by the combined mechanism. In the very flexible frame the structure reaches its maximum load before a plastic collapse mechanism forms. In this case the structure becomes unstable when the third hinge forms at A and the structure buckles, as can be seen by the very rapidly increasing deflection.

Wood [16] has explained this type of behaviour. Just as the column in the previous section had an elastic critical load, so too does the frame. (Horne and Merchant [14] have presented a procedure for determining its magnitude.) Usually this load is much greater than the plastic collapse load, as can be seen in table 6.4. However, each time a plastic hinge forms, the stiffness of the frame

l/r	Collapse load factor	λ_e modified frame	λ_e original frame	λ_c (simple plastic analysis)	λ_R equation 6.7
100	49.10	400	2124	50.0	48.9
200	48.25	200	1062	50.0	47.8
500	45.93	80	424	50.0	44.7
100*	46.67	200	531	50.0	45.7
200*	45.13	100	265	50.0	42.1
500*	36.60	7.6	106	50.0	34.0

* Frame loaded as in figure 6.11b

is reduced. The elastic critical load of the frame is now the load for a modified frame with a frictionless hinge at the plastic hinge position. The frame and elastic critical load must be modified successively as each hinge forms. Table 6.4 also shows the elastic critical load of the modified frame when the third plastic hinge has formed. In the case of the flexible frame ($l/r = 500$) that elastic critical load is smaller than the applied load so that buckling must occur.

6.3.3 The Rankine–Merchant Load Factor

The effects of axial force and deflection are rather disturbing. Except perhaps for single-storey frames, it is not enough just to calculate the simple plastic collapse load. It would also be a violation of the elegance and simplicity of the plastic methods to resort to the non-linear computer analysis.

Very stiff structures collapse at the simple plastic collapse load, while very flexible ones buckle at the elastic critical load. In general, these loads can be found without too much difficulty. Merchant devised a means of approximating the true collapse load factor, from the simple plastic collapse and elastic critical loads, based on the Rankine amplification factor used in strut analysis. (The simple plastic collapse load is obtained by the methods described in chapters 3 and 4, which ignore deflections and axial loads.) This approximation, called the *Rankine–Merchant load factor*, λ_R, is defined by the equation

$$\frac{1}{\lambda_R} = \frac{1}{\lambda_c} + \frac{1}{\lambda_e} \tag{6.7}$$

where

$$\lambda_c = \text{simple plastic collapse load factor}$$

$$\lambda_e = \text{elastic critical load factor}$$

It is plotted in figure 6.14, with the failure loads of several frames tested by Low. The Rankine–Merchant load factor gives in every case a safe approximation of the observed collapse load factor. The Rankine–Merchant load factors for the portal frames in the previous section are given in table 6.3. In every case the Rankine–Merchant approximation is close to, but lower than, the theoretical collapse load factor.

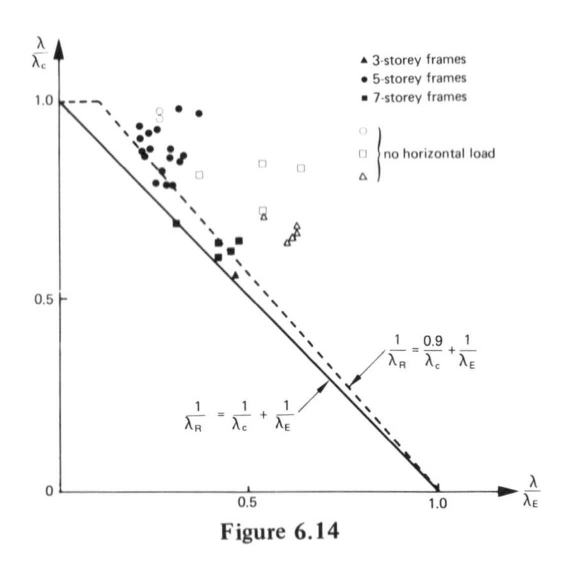

Figure 6.14

The apparent conservatism of λ_R as shown in figure 6.14 is partly due to strain hardening during the testing. Wood [17] has suggested the following modification to equation 6.7 to get a better approximation

$$\lambda_R = \lambda_c \qquad \text{when } \frac{\lambda_e}{\lambda_c} > 10$$

$$\frac{1}{\lambda_R} = \frac{0.9}{\lambda_c} + \frac{1}{\lambda_e} \qquad \text{when } 10 > \frac{\lambda_e}{\lambda_c} > 4 \tag{6.8}$$

This is shown by the broken line in figure 6.14. As can be seen, it agrees more closely with the experimental results than the Rankine–Merchant value. When $\lambda_e/\lambda_c < 4$, Wood suggests that the simple analysis is insufficient. It seems likely that Wood's modified equation will be included in the new British code for steel design.

6.4 SUMMARY

This chapter has been concerned with two things: the calculation of deflections at the point of collapse, and the effect that those deflections (and the axial forces) can have on the collapse of the structure.

It was shown initially why it is important in some cases to know the magnitude of deflections before collapse, because limiting those deflections might be more critical than ensuring that the structure is sufficiently strong.

The slope deflection method of calculating these deflections was considered next. The various stages of the procedure are

(1) Determine the collapse mechanism, the corresponding load factor, BMD and end moments (including fixed end moments) of each member of the structure.

(2) Write down the slope deflection equations for each member.

(3) Obtain relationships between the various unknown deflections by considering continuity at every member connection which is shown by the BMD to be elastic. Calculate the deflections, assuming that each plastic hinge in turn is the last one to form.

(4) Choose the last hinge to form and the corresponding set of deflections by using the displacement theorem.

The final part of the chapter was an examination of the non-linear behaviour which results from the axial forces in the members. It was shown that the effect is a reduction in the collapse load factor of the structure, the reduction depending on the stiffness (as measured by slenderness ratio) of the structure: the lower the stiffness, the greater the deflections and reduction in collapse load factor. In single-storey frames the reduction is not usually significant for a structure with practical values of slenderness ratio in the columns. In multi-storey frames the reduction can be more serious, leading to premature buckling before plastic collapse can occur. The Rankine–Merchant load factor was shown to give a good estimate of the reduced collapse load factor, although it appears to be rather conservative when compared to test results. Wood's modification agrees more closely with these results.

6.5 PROBLEMS

6.1 A fixed end beam, span L, carries a vertical load W at a distance $L/3$ from the left hand support. Assuming a constant M_p and EI for the beam determine the vertical deflection at collapse under the load.

6.2 A propped cantilever, span L, carries a UDL w per unit length. Determine the vertical deflection, at collapse, at the plastic hinge near the centre of the span. Assume M_p and EI are constant.

6.3 Find the horizontal and vertical deflections, at collapse, at mid-span of the portal frame shown in figure 6.15.

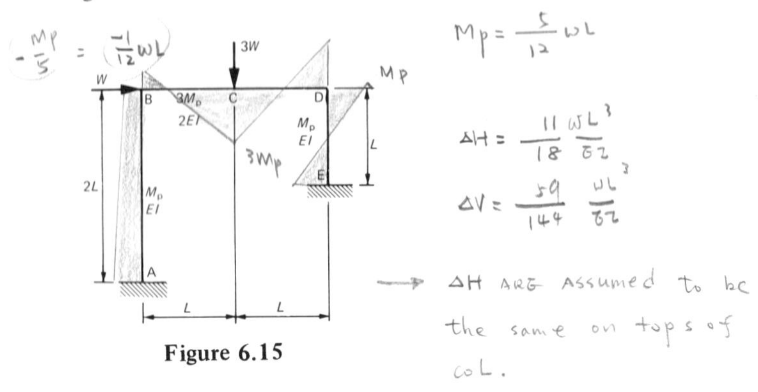

Figure 6.15

6.4 The sway deflection at collapse of the frame shown in figure 6.16 is the same at the top of both columns. Determine its magnitude, given that EI = 10 000 kN m^2 and M_p = 100 kN m.

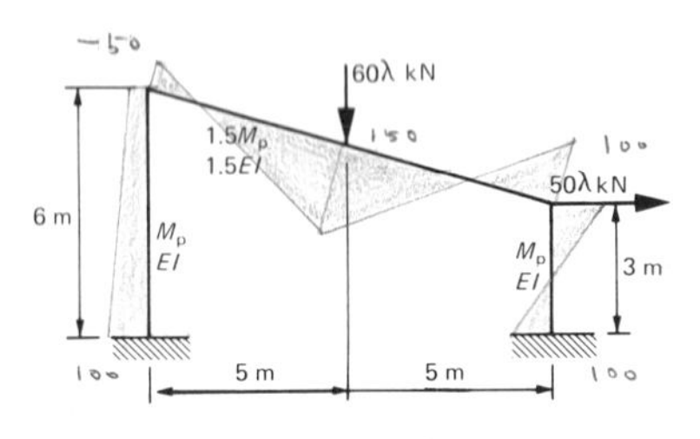

Figure 6.16

6.5 The pitched portal frame shown in figure 6.17 fails by the combined mechanism when $\lambda W = 5M_p/22$. Determine the horizontal and vertical deflections of the ridge at the point of collapse.

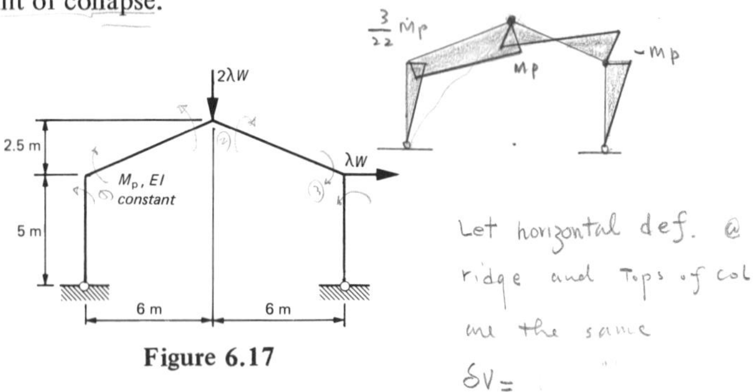

Figure 6.17

6.6 The two-bay frame shown in figure 6.18a has a load factor of 2.0 against collapse. The BMD at collapse and the plastic hinges are shown in figure 6.18b.

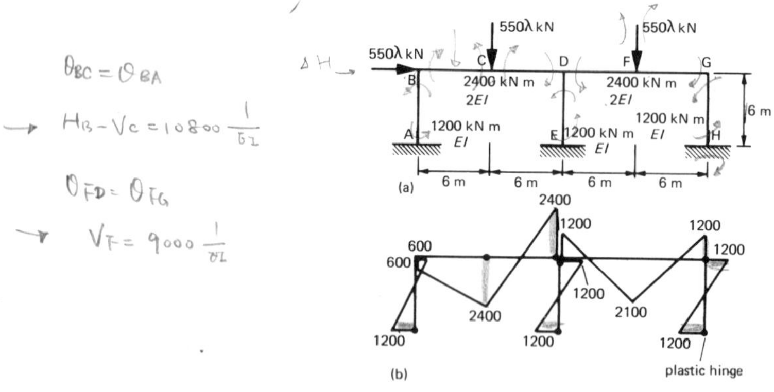

Figure 6.18

Calculate the horizontal deflection at B and the vertical deflections at C and F, at the point of collapse. (Hint: there are no plastic hinges at B and F).

6.7 A fixed end beam, length L, carries an axial force P and a UDL of intensity $P/2L$ per unit length. The beam has a rectangular cross-section, width b and depth d, so that

$$M_p = \frac{bd^2}{4}\sigma_y \qquad P_p = bd\sigma_y$$

and the reduced plastic moment due to axial force, M_p', is given by

$$M_p' = M_p(1 - n^2)$$

where $n = P/P_p$. Assuming that $d = L/24$

(a) calculate the simple plastic collapse load (in terms of P), using the free and reactant BMD method.

(b) Show that the reduced collapse load when there is a finite deflection Δ at the midspan hinge is given by solving the equation

$$\frac{\Delta}{L} = \frac{1}{48}\left(\frac{1}{n} - 3 - n\right)$$

(c) the elastic deflection at midspan is given by

$$\frac{\Delta}{L} = \frac{n}{32(3 - n)}$$

Find an estimate of the true collapse load of the beam (in terms of P).

(d) find the Rankine–Merchant load (in terms of P). The elastic critical load can be found from the equation in (c).

7 APPLICATION OF PLASTIC METHODS TO REINFORCED CONCRETE STRUCTURES

7.1 INTRODUCTION

At first glance concrete structures bear little resemblance to steel ones. Unexpectedly, reinforced concrete (RC) beams do, in certain circumstances, act similarly to steel beams, because of the characteristics of the steel reinforcement which control the behaviour of the beam. The maximum BM that a section can carry, usually called the *moment of resistance* of the section, is calculated in a similar way to the plastic moment of a steel beam. Many tests on RC beams have shown that the calculated moment of resistance is very close to the experimental value, confirming the applicability of the theory. Section 7.2 looks in some detail at the analysis of reinforced concrete sections carrying BMs only.

Unfortunately, there are problems in trying to use conventional plastic methods on RC frames. The moment of resistance analysis shows that RC sections may not have much capacity for plastic rotation. It is that capacity which is essential to achieve the redistribution of moments required by the plastic methods. Section 7.3 contrasts what can happen if plastic rotation capacity is exceeded.

Despite these problems, modern British RC design practice in particular attempts to make use of and benefit from plastic theory. The way that this is achieved is described in section 7.4.

It must be emphasised that this chapter is concerned only with the application of plastic theory to reinforced concrete frames. The reader should consult the appropriate texts on other aspects of reinforced concrete design. [18, 19]

7.2 THE BEHAVIOUR OF REINFORCED CONCRETE IN BENDING

7.2.1 Assumptions

It was shown in chapter 2 how plastic hinges form in steel members. RC beams can be considered in a similar manner, although the analysis is more complicated. The following assumptions are necessary.

(1) Plane sections remain plane so that longitudinal strain is directly

proportional to distance from the axis of zero strain. This has been shown to be substantially true up to failure by several researchers. [20, 21]

(2) Concrete has no strength in tension.

(3) The stress–strain curve for concrete in compression, due to bending, is similar in shape to that for direct compression. The maximum stress is smaller because the concrete is restrained differently. Hognestad *et al.* [21] suggested the curve shown in figure 7.1. The maximum strain in the concrete depends on the compressive strength of the concrete, [21] as does the maximum stress. The stress reaches a maximum value and then falls off before the concrete fails.

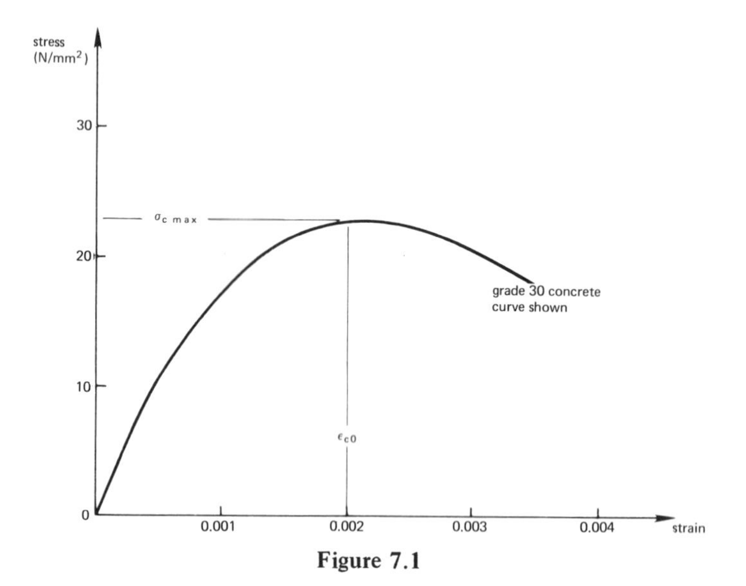

Figure 7.1

7.2.2 Beams with Tension Reinforcement Only

A typical rectangular concrete beam is shown in section and elevation in figure 7.2 at some intermediate stage of loading. In addition to the horizontal tension bars shown in the section there are also vertical links which help resist shear. The vertical cracks in the concrete tend to occur close to these links. The BM is

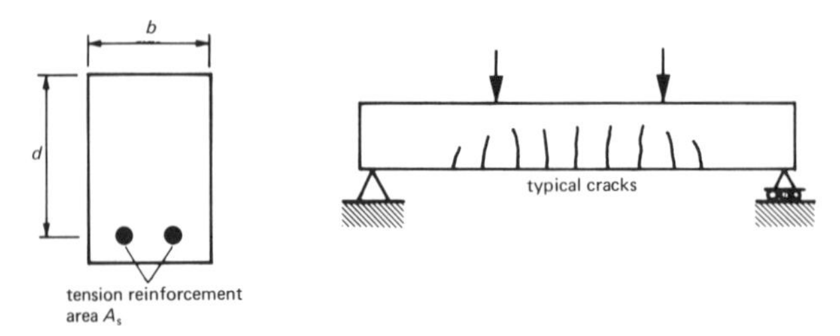

Figure 7.2

constant between the two vertical applied loads so that theoretically the strain distribution should be identical at every section in that length. This is reasonably correct in the concrete in compression, but obviously not so in the tension region because of the cracks. Tests have shown [22] that the strain in the steel varies, being a maximum at the cracks. The variation is due to bonding of the concrete and steel between the cracks. The variation is ignored in calculations, but the bond should not be forgotten, without it there would be no reinforced concrete, merely separate masses of steel rod and concrete.

The analysis has been carried out for a beam of grade 30 concrete (28 day cube strength, 30 N/mm^2) for various percentages ($100A_s/bd$) of mild and high-tensile steel reinforcement. The materials data are given in tables 7.1 and 7.2. The concrete stress–strain curve was modelled by the two parabolas

$$\sigma = 0.76\, f_{cu} \frac{\epsilon}{\epsilon_{co}} \left(2 - \frac{\epsilon}{\epsilon_{co}}\right) \qquad \epsilon < \epsilon_{co} \qquad (7.1)$$

$$\sigma = 0.76\, (f_{cu} - 2.85 \times 10^6 \qquad (\epsilon - \epsilon_{co})^2)\ \epsilon > \epsilon_{co} \qquad (7.2)$$

Since the strain distribution is known (from the first assumption) the stresses in the steel and concrete can be found from their respective stress–strain curves.

Table 7.1 Concrete properties

cube strength (N/mm^2)	30
σ_c max (N/mm^2)	22.8
ϵ_c at failure	0.0035
ϵ_{co}	0.002

Table 7.2 Reinforcement properties

	Mild steel	High tensile
Yield stress (N/mm^2)	250	410
E (kN/mm^2)	210	205
Strain at yield	0.00119	0.002
Strain hardening strain	0.015	0.007
E_{SH} (kN/mm^2)	8.4	8.2

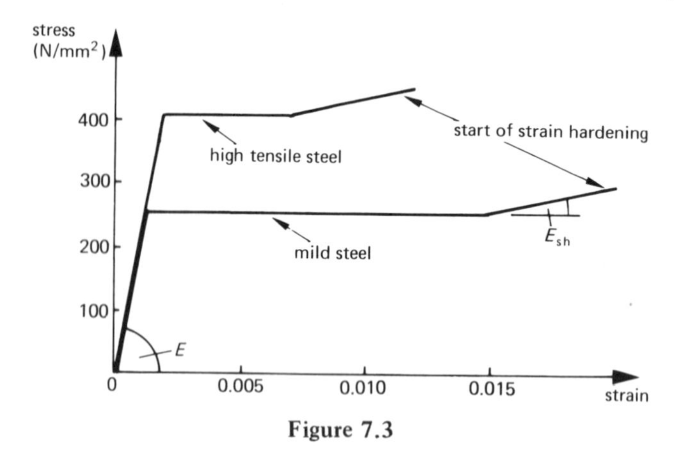

Figure 7.3

(Typical curves for reinforcing steel are shown in figure 7.3.) Stress distributions at various stages are shown in figure 7.4. The section can be analysed at each stage in the same way as the steel section in chapter 2. The complication is that the zero-strain axis cannot usually be found directly. The steps in the analysis are as follows.

(1) Choose concrete strain and initial value for x (which defines the axis of zero strain, as in figure 7.4).

(2) Find the compressive force C in the concrete, and the steel tension T.

(3) For horizontal equilibrium, $C = T$. If the difference between them is more than 0.1 per cent, adjust x and return to (2).

(4) Find the moment of C and T about the axis of zero strain; the sum gives the BM which causes the chosen concrete strain. The sum of the maximum concrete and steel strains divided by d gives the corresponding curvature (see chapter 2).

The analysis does not always converge rapidly and a small computer is really necessary.

The results, in the form of moment–curvature curves, are shown in figure 7.5. The curves are interesting. At low reinforcement percentages the curves are very

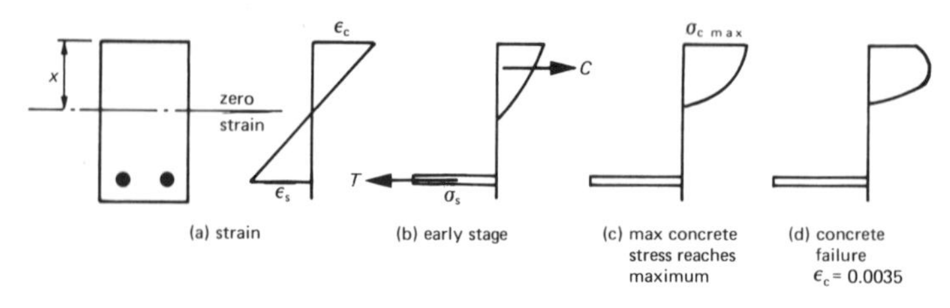

Figure 7.4

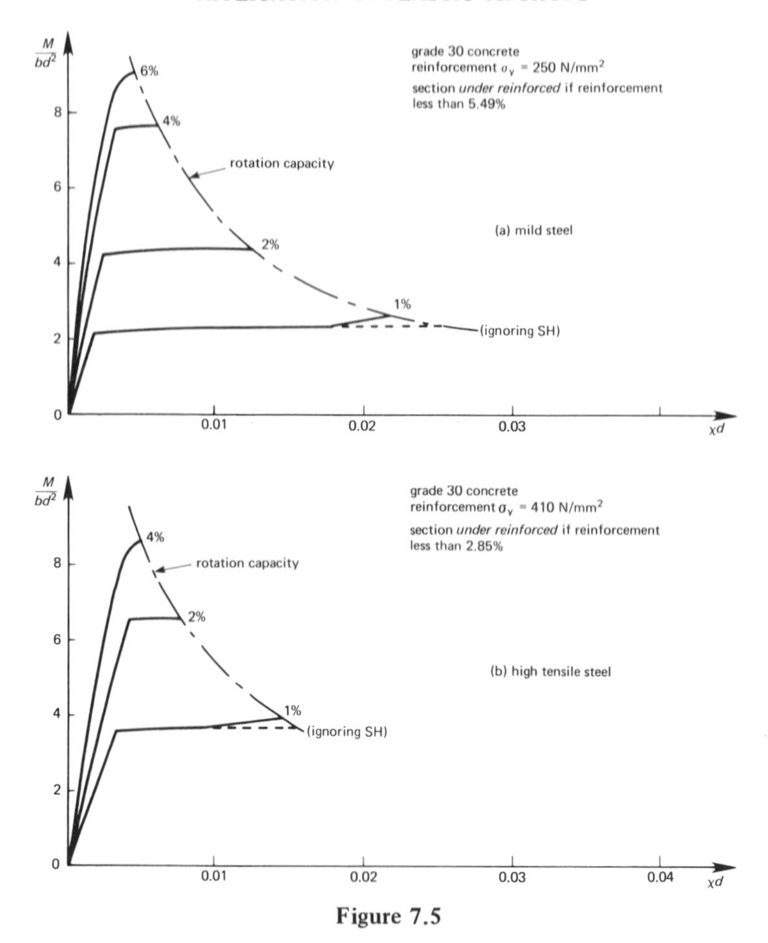

Figure 7.5

similar to those for steel beams, an initial almost straight line showing rapid increase of moment with only a small increase in curvature (corresponding to the elastic region), which bends over to give large curvature with only a very small increase in moment (corresponding to plastic rotation). As the reinforcement percentage increases the maximum moment becomes proportionally larger, but there is also a decrease in maximum curvature.

The rapid curvature increases with low reinforcement are caused by the steel yielding before the concrete reaches its maximum stress. Such behaviour is called *under-reinforced.* The large curvatures give rise to large increases in deflection so that there is warning of impending failure, which eventually occurs by the concrete crushing.

At the higher percentages failure is very sudden and often explosive, with no large deflection increase prior to failure. The concrete crushes before the steel yields. This is called *over-reinforced* behaviour.

Under-reinforced sections are effectively controlled by the reinforcement and behave like steel beams. Ultimately the concrete fails when its maximum strain reaches a limiting value (taken as 0.0035 in the example), which is considerably less than the maximum possible steel strain, so that the capacity for plastic rotation is very limited compared to steel beams. Over-reinforced sections are controlled by the brittle concrete and have no plastic rotation capacity.

At very low percentages there is a pronounced increase in moment as well as curvature just before failure due to the onset of strain hardening in the reinforcement.

There is one other major difference between under- and over-reinforced behaviour. Figure 7.6 shows the position of the axis of zero strain at failure for various reinforcement percentages. Increasing the reinforcement causes the axis to drop. Limiting the position of the axis of zero strain is a convenient way of preventing the problems of over-reinforced behaviour.

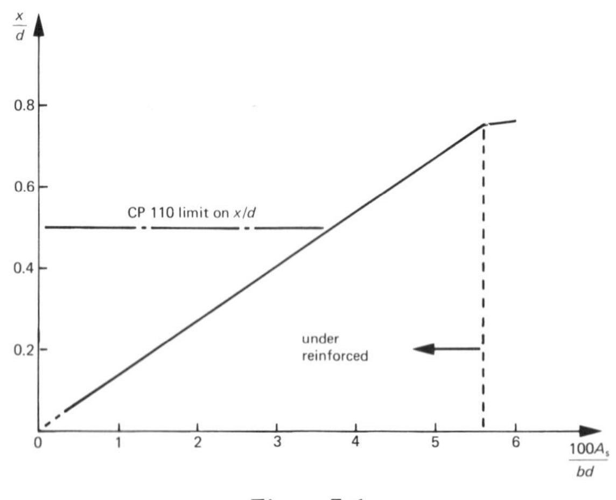

Figure 7.6

7.2.3 Beams with Tension and Compression Reinforcement

Virtually all RC beams have some compression reinforcement. Often it will only be a nominal quantity to provide a framework on which to place the shear links, but it may be provided to boost the compressive strength of the concrete. A typical cross-section is shown in figure 7.7. The behaviour can be analysed in the same manner as before, but including the extra force in the compression reinforcement. The results of such an analysis are shown in figure 7.8, assuming that $d_1/d = 0.1$.

The results show that as well as increasing the moment capacity of the section, compression reinforcement also increases, dramatically in some cases, the plastic rotation capacity of the section.

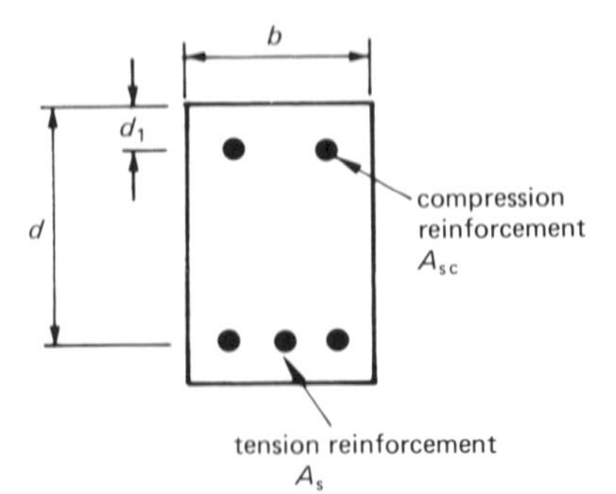

Figure 7.7

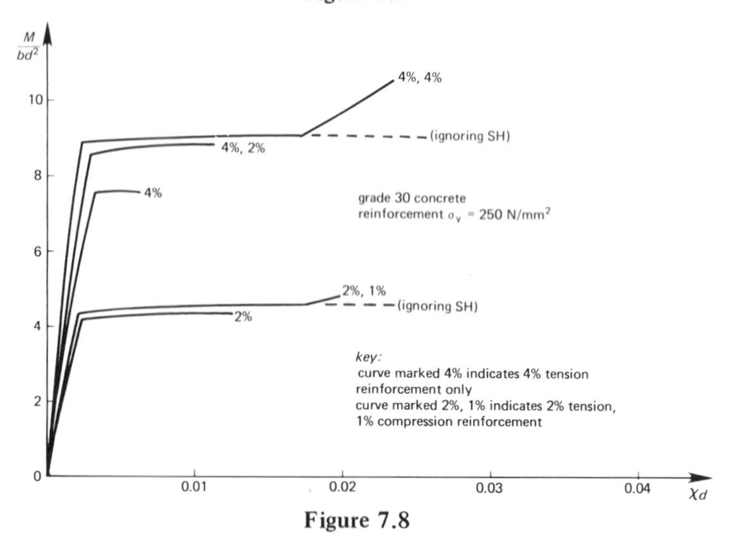

Figure 7.8

7.2.4 Summary of Moment–Curvature Relationships

The analyses presented in the two previous sections have been summarised in the two graphs of figure 7.9. The graphs are plots of the ratio of moment to ultimate moment against the ratio of curvature to ultimate curvature. Figure 7.9a compares results for beams containing various percentages of tension reinforcement and a typical steel beam. It shows clearly how the proportion of curvature available for plastic rotation drops as the reinforcement increases, but it also shows that the steel beam has very much more plastic rotation capacity. Figure 7.9b shows the beneficial effect of the compression reinforcement in increasing plastic rotation.

The results vary with different concrete strengths, but there is always clearly defined under- and over-reinforced failure. Table 7.3 shows the reinforcement percentage and the position of the zero strain axis (expressed x/d) at the transition from under to over-reinforced behaviour for various concrete strengths and types of reinforcement.

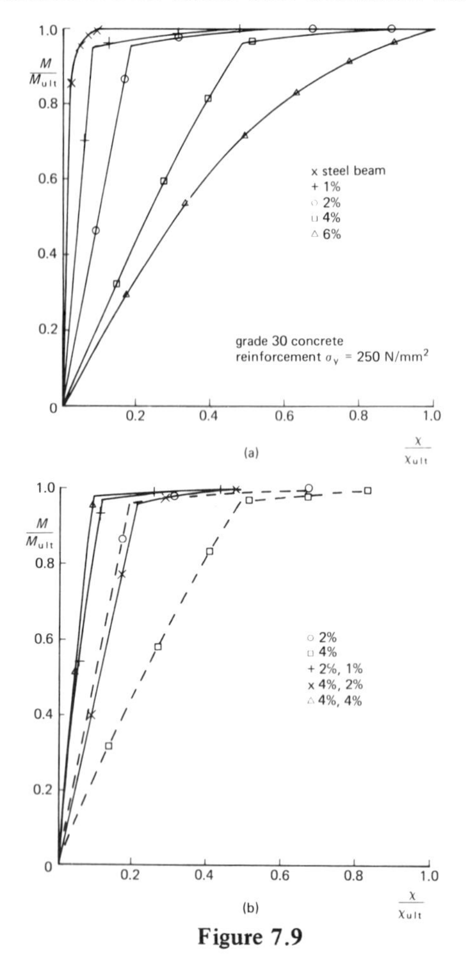

Figure 7.9

Table 7.3

Concrete Grade	Mild Steel Reinforcement		High Tensile Reinforcement	
	%	x/d	%	x/d
20	4.11	0.75	2.14	0.64
25	4.85	0.75	2.53	0.64
30	5.49	0.75	2.85	0.64
40	6.63	0.74	3.43	0.62
50	7.62	0.72	3.91	0.61

The results used in this chapter are theoretical, but experiments confirm that the theory accurately describes the behaviour of reinforced concrete sections.

The analysis used here can be applied to any shape of concrete beam and even to columns carrying moments and axial forces.

7.3 WHAT HAPPENS IF THERE IS INSUFFICIENT PLASTIC ROTATION CAPACITY?

Consider the continuous RC beam shown in figure 7.10a. An elastic analysis (BMD in figure 7.10b) shows that the largest BM is at the central support. Assuming that moment equals the ultimate moment of the reinforced concrete the subsequent behaviour can follow two routes depending on the amount of reinforcement in the beam. Parts c and d of figure 7.10 show the alternatives. If

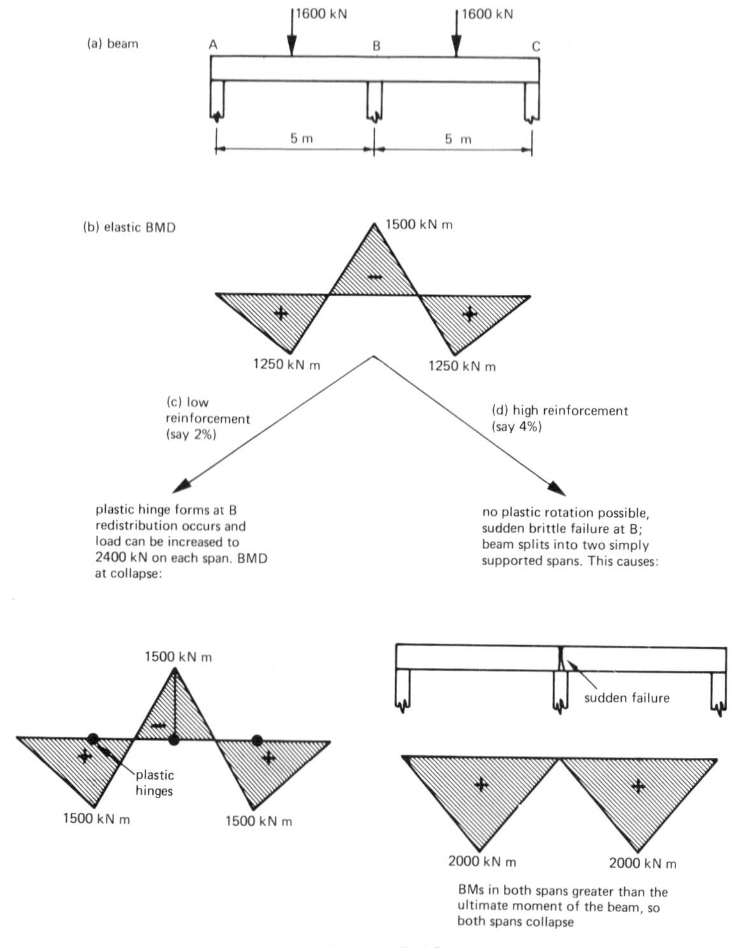

Figure 7.10

there is only a small percentage of reinforcement in the beam, the reinforced concrete has sufficient plastic rotation capacity for full redistribution of the bending moments and the beam is able to withstand more load before collapse occurs. However, when there is a high percentage of reinforcement, brittle rupture occurs at the point of highest BM and the beam splits into two simply supported spans. The resulting BMs in each span are considerably greater than the moment of resistance of the beam and each span collapses immediately.

The differences in behaviour are very marked, but they represent the two extremes. In general some plastic rotation is possible so that there can be some redistribution of the bending moments. Once the rotation capacity is reached sudden catastrophic failure will occur. This illustrates the difficulty of designing RC structures by plastic methods. It is not sufficient to check that the structure has sufficient resistance against collapse (which is checked by simple plastic calculations), it is also necessary to check that the plastic rotation at each hinge is within the plastic rotation capacity of the section so that the mechanism can develop. In the simple plastic methods, for example virtual work, the magnitude of the rotations cannot be found.

As the model portal frames in figure 7.11 show, reinforced concrete frames can be designed to develop full collapse mechanisms. Indeed, Baker [23, 24] has put forward an elegant plastic design method for reinforced concrete which tests

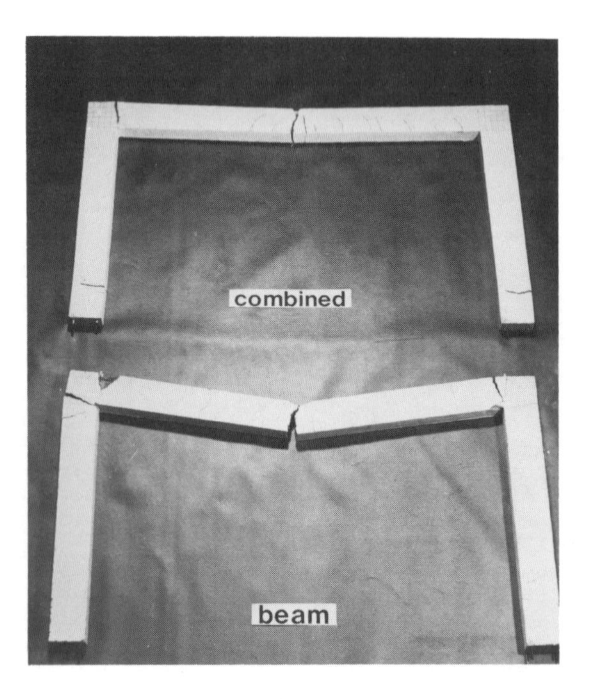

Figure 7.11 Collapse mechanisms for model RC portal frames

the magnitude of the rotations. It is obviously more complicated than the simple methods. Other more complex methods also exist [25]. However, the problem of limited rotation capacity, despite the hidden benefit of the compression reinforcement, has prevented the adoption of 'true' plastic methods of design for reinforced concrete frames.

7.4 THE COMPROMISE ADOPTED IN CODES OF PRACTICE

Current British design practice for reinforced concrete is outlined in CP110. [5] The code allows a highly modified form of plastic design, without actually admitting that it does. Other codes adopt a similar, but less sophisticated, approach.

CP 110 also embraces the full limit state philosophy. When moments of resistance are calculated, partial safety factors are applied to the strengths of the concrete and reinforcement. The concrete stress block given in figure 7.1 was based on the results of experimental research, and is too complicated for design work. CP 110 allows two alternative blocks for finding the moment of resistance. The simplified stress blocks are compared with the one used above (allowing for the partial safety factors) in figure 7.12. All three give surprisingly similar moments of resistance as shown in table 7.4.

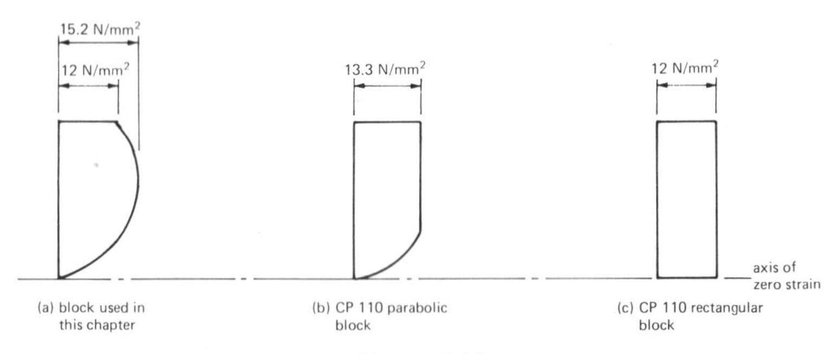

Figure 7.12

One of the main conclusions from sections 7.2 and 7.3 was that over-reinforced sections are dangerous and must never be used. CP 110 prevents over-reinforcement in two ways. Firstly it limits the depth to the axis of zero strain to half the effective depth ($x/d \leqslant 0.5$) and secondly it limits the reinforcement to a maximum of 4 per cent. As was shown in table 7.3 and figure 7.5 these limits will ensure under-reinforced behaviour.

Thus far the code has met all the requirements of plastic theory in that it has ensured under-reinforced sections with some plastic rotation capacity, although that capacity will be very limited for some sections. As explained in section 7.3 'true' plastic design cannot be permitted for finding the magnitude of the

Table 7.4 Comparison of concrete stress blocks

Grade 30 Concrete	Mild steel reinforcement		
	M/bd^2		
100 As/bd	Block used in this chapter	CP 100 parabolic block	CP 100 rectangular block
1%	2.00	1.99	1.97
2%	3.64	3.63	3.56
x/d = 0.5 when	4.73 (2.8% reinforcement)	4.63 (2.8% reinforcement)	4.50 (2.8% reinforcement)

moments of resistance. Instead CP 110 gives the following rules for determining the moments of resistance.

(1) Use *elastic* analysis to find the BM distributions, using *factored collapse* loads. The elastic analysis will give a set of reactant BMs at the end of each member. Intermediate moments can be found by combining the free BMs with the reactant moments.

(2) These moments can now be redistributed by using the following rules:

(a) there is no limit when moments of resistance are increased by redistributions.

(b) if redistribution causes a reduction, the amount of reduction is limited by a reduction factor β_{red}

$$\beta_{red} = \frac{\text{reduction in } M_r}{\text{largest elastic moment in the member}}$$

$$\beta_{red} \not> 0.3 \text{ or } 0.6 - \frac{x}{d} \tag{7.3}$$

x/d = neutral axis factor for the section resisting the *reduced* moment. (This implies that a maximum reduction of 30 per cent is possible if the resulting section is very under-reinforced (see table 7.3). As the percentage of reinforcement increases, so the neutral axis factor, x/d, increases and the plastic rotation capacity and amount of redistribution decreases).

In beams x/d_{max} is 0.5 so that the minimum reduction is 10 per cent. Columns have to carry axial force as well as moments and CP 110 allows x/d to take any value. Unless the axial force is small it is unlikely that any significant reduction can be allowed in column moments.

(3) The moment of resistance at any section must not be less than 70 per cent of the maximum elastic moment at that section.

A simple example illustrates the importance of this rule. Figure 7.13 shows the BMD for a beam with fixed ends. Using rule 1, the BMs are found by elastic analysis using the collapse loads. At normal working loads the BMs are also elastic and a proportion of the 'collapse' elastic moments, as shown. Redistribution of the moments is achieved by reducing the reactant moments at each end. The redistributed moments are the required moments of resistance, but they also define where the reinforcement is placed. Remembering that the reinforcement

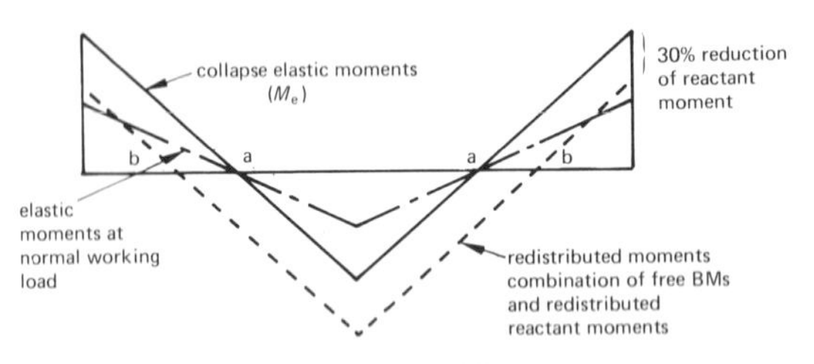

Figure 7.13

resists the horizontal tension due to bending, the reinforcement should be placed as in figure 7.14. The problem is that the switch from top to bottom reinforcement is made at point b, the point of contraflexure on the redistributed BMD. Under normal working loads reinforcement is still required and in the top of the beam it should extend to point a, as shown by the broken line in figure 7.13. The resulting cracks would not give rise to much confidence in the beam! Rule 3 effectively prevents this problem, it would give rise to the pattern of moments of resistance in figure 7.15.

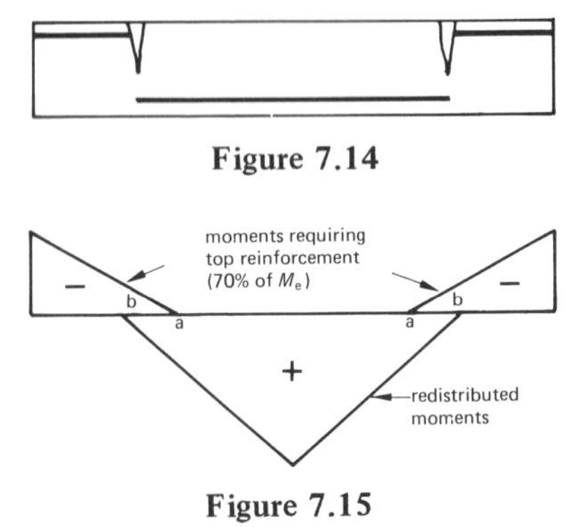

Figure 7.14

Figure 7.15

CP 110 recognises that full redistribution is not always essential in a design. If the redistribution is less than 10 per cent everywhere some simplified formulae are provided for calculating moments of resistance.

There are two main strategies when redistributing the moments. If the member has a rectangular cross-section, the aim must be to equalise as much as possible the maximum hogging and sagging moments and thus avoid difficulties in positioning the reinforcement. If the beam has a T or L cross-section the hogging moments must be made as small as possible, since the hogging moment of resistance is smaller than the sagging moment of resistance because of the smaller area of concrete in compression. The redistribution process is best illustrated with an example.

Figure 7.16 shows the complete example. It is assumed that the continuous beam has a T-section and carries a maximum factored load of 200 kN m and a minimum load of 50 kN m. The two worst load combinations are shown. The first causes the largest hogging moment at the central support, the second the largest sagging moment. The 'elastic' moments were found using moment distribution.

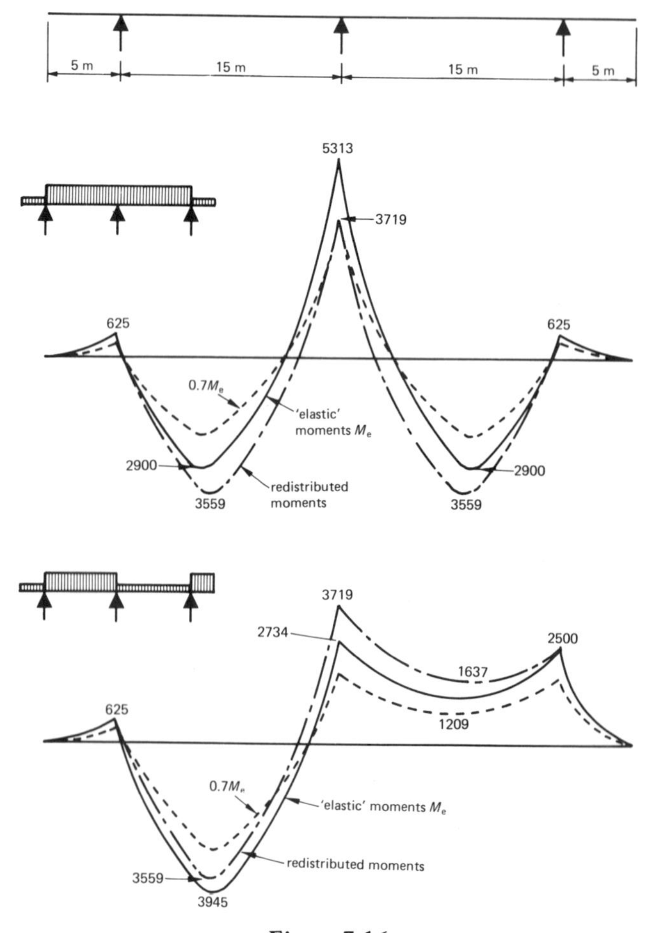

Figure 7.16

The object of the redistribution is to reduce as much as possible the hogging moments. The central hogging moment in load case 1 is reduced by the full 30 per cent to 3719 kN m. The moments at the ends of the cantilevers cannot be changed since they are in equilibrium with load on the cantilevers. The free BMs are then hung on the new reactant moments. These redistributed moments and 70 per cent of the original moments are plotted.

In load case 2 the central hogging moment can be increased to 3719 kN m since that moment of resistance is provided for load case 1. The resulting sagging moment in the left-hand span is 3559 kN m (reduced from 3945 kN m). The reduction of 386 kN m gives a β_{red} of 0.072 (= 386/5313).

The worst BMs from the two load cases can be combined to give the design BM envelope. Remember that in some parts of the beam $0.7M_e$ will be more critical than the redistributed moments. The envelopes for before and after redistribution are shown in figure 7.17. (It should be realised that there is a mirror image of load case 2 when working out the envelope). The benefit of the redistribution is obvious. The maximum hogging moment is considerably reduced, and almost accidentally in this case, so is the maximum sagging moment. The cut-off points (where sagging or hogging moments are zero) can be found by considering the geometry of the BMDs (see section 4.3.3) or by drawing the BMDs accurately and measuring off the dimensions.

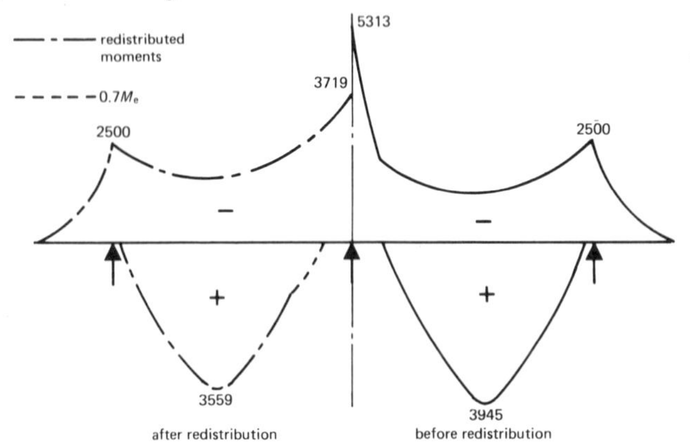

Figure 7.17

There is one further check which must be carried out. At the central support β_{red} = 0.3, which means that the maximum value of the neutral axis factor x/d at that point is 0.3 (see equation 7.1). At the point of maximum sagging moment β_{red} = 0.072, but x/d can only be 0.5, not 0.528 as might be expected, because CP 110 limits the maximum value to 0.5 to ensure under-reinforced behaviour.

7.5 SUMMARY

This chapter has shown the problems of applying plastic methods to the design of RC structures. Under-reinforced members are capable of plastic rotation, but the capacity for rotation is limited compared to steel beams. This has made it impossible to apply simple plastic theory. CP 110 has ensured under-reinforced sections by limiting the amount of reinforcement and the position of the neutral axis. The process of moment redistribution is an attempt to gain the benefits of the plastic methods and is subtle enough to allow for the reduction in rotation capacity as the amount of reinforcement is increased.

There is one type of concrete structure which is usually very under-reinforced. Concrete slabs generally only have small percentages of reinforcement and are designed by plastic methods. These methods are described in the next chapter.

7.6 PROBLEMS

7.1 A fixed end RC beam, span L, carries a UDL w per unit length at collapse. Redistribute the bending moments to achieve equal moments of resistance at midspan and the supports. What is the magnitude of the moment of resistance? Determine the cut-off points for top and bottom reinforcement.

7.2 Figure 7.18 shows an RC T-beam, carrying a collapse load of 300 kN m (including self-weight), and the BMD based on elastic analysis. Redistribute the moments to obtain a more satisfactory BM distribution. Plot this on a diagram giving the most important moment values and dimensions.

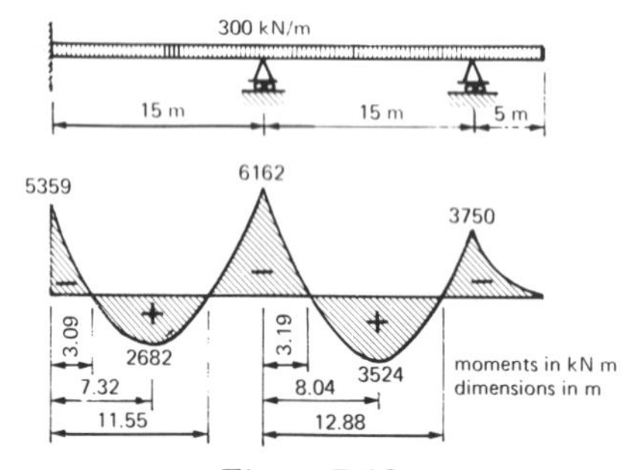

Figure 7.18

7.3 Repeat problem 7.2 for the beam and BMD shown in figure 7.19.

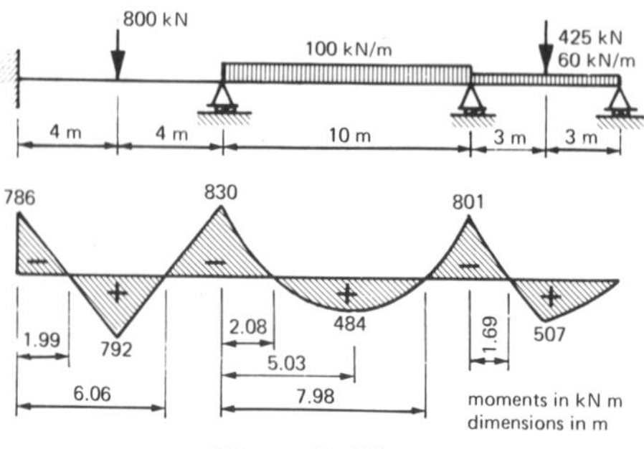

Figure 7.19

7.4 A continuous beam is shown in figure 7.20 together with the BMDs for the three most critical load cases. Use moment redistribution to even out the maximum hogging and sagging moments. Summarise the results by drawing the envelope of redistributed moments.

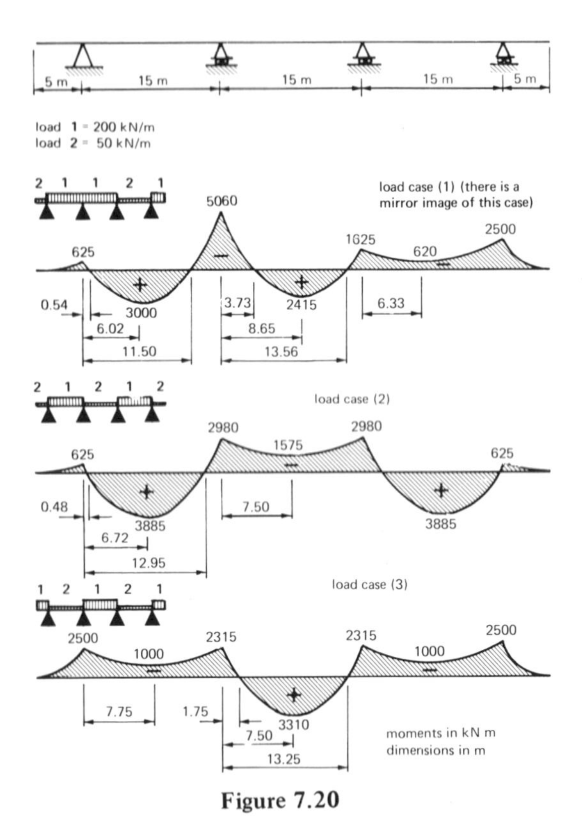

Figure 7.20

8 YIELD LINE ANALYSIS AND THE HILLERBORG STRIP METHOD FOR REINFORCED CONCRETE SLABS

8.1 INTRODUCTION

The previous chapter showed that plastic theory is usually rather difficult to apply to RC structures because of the limited plastic rotation capacity of concrete sections in bending. However, concrete slabs are almost always very under-reinforced. It is unusual to have sections with more than 1 per cent of reinforcement. Consequently slabs have considerable plastic rotation capacity and can be successfully analysed or designed by plastic methods.

This chapter presents two methods which are now widely used. Yield line theory is the more well-known of the two. It has similarities, as shown here, to the methods of finding the collapse loads of steel structures. The Hillerborg strip method is a novel approach to slab design, which is simple and convenient to use.

8.2 YIELD LINE THEORY

8.2.1 An Experimental Basis for Yield Line Analysis

The pioneer of yield line theory was K. W. Johansen, who published his doctoral thesis on the subject in 1943. [26] Many researchers have extended Johansen's ideas, perhaps further than he envisaged or desired. Their intention has been to tie yield line theory in with more classical (and rigorous) plastic theory.

It is easiest for the beginner to approach the theory via the results of tests on concrete slabs. The photographs in figures 8.1, 8.5 and 8.16 show the top and bottom faces of three model slabs which have been tested with a uniform load over the slab surface (UDL) and various edge conditions.

Figure 8.1 shows a square slab with reinforcement parallel to the sides of the plate. The edges were supported to prevent vertical movement but offered minimal resistance to rotation (called simply supported edges). On the tension side of the slab (the underside of a floor slab but called the top in the figure

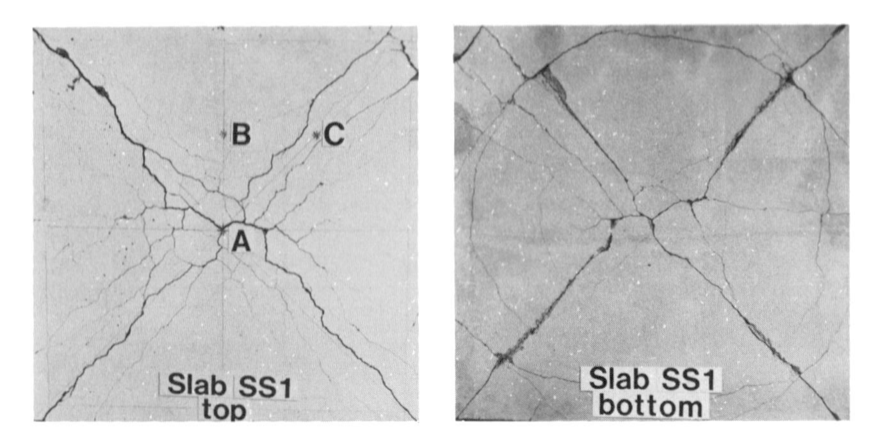

Figure 8.1 Test on square simply supported slab

because the test rig applies an upward load to the slab) there are many cracks, especially near the centre of the slab. This web of cracks forms at a relatively low load, but as the load increases only a few of the cracks become large. In this particular case the cracks roughly on the diagonals have opened out. The other (compression) face of the slab is relatively unmarked apart from crushing of the concrete which coincides with the large cracks on the tension face. This combination of large cracks and concrete crushing is exactly what occurs when an under-reinforced beam reaches its moment of resistance after plastic rotation. A yield line is an idealisation of this situation. It is perhaps simplest to visualise the yield line as a series of beam sections side by side, with each section undergoing plastic rotation at its moment of resistance.

During the test dial gauges were placed at the points marked A, B and C to measure the vertical deflection of the slab. The load deflection curves for the three gauges are shown in figure 8.2. The curve for the central deflection shows a change in slope at a low load (when cracking first starts) and then a gradual reduction in slope as the yield lines spread through the slab. The slope does not drop continuously to zero (indicating collapse) as might be expected. This is a point which will be taken up later (see section 8.2.3).

The curves for points B and C diverge initially but become practically identical at higher loads. When cross-sections are plotted at different loads through the centre line through the points A and B, as in figure 8.3, it can be seen that at higher loads the cross-section is almost a straight line. This, and the identical deflections at B and C, shows that at loads near to the collapse load the regions between the yield lines are practically flat.

This experimental behaviour leads to a relatively simple model of the collapse of the slab. The collapse mechanism consists of two yield lines running the full length of the diagonals with rigid triangles of slab between the yield lines

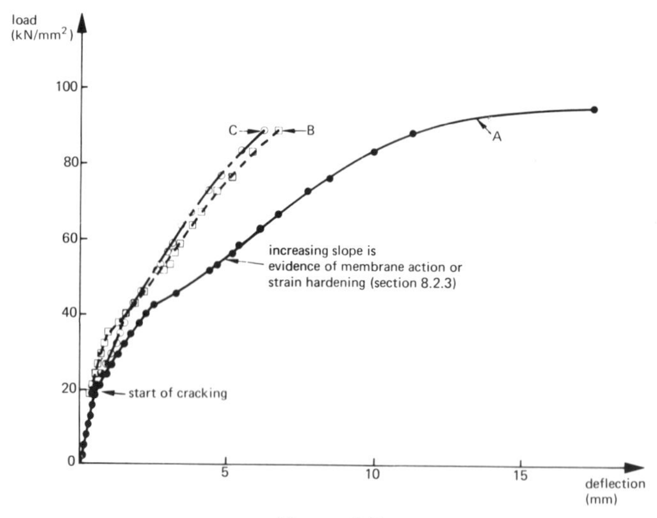

Figure 8.2

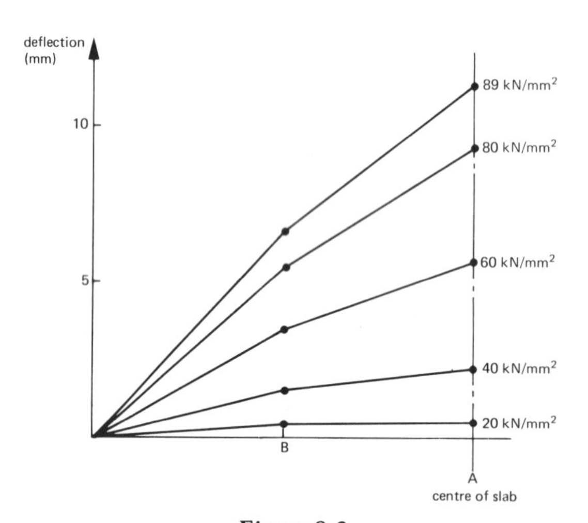

Figure 8.3

(figure 8.4). The edges of the slab and the yield lines are axes of rotation for the rigid regions of the slab, which means that the plastic rotation of the yield lines is the only deformation of the slab. Chapters 3, 4 and 5 were concerned basically with the analysis of collapse mechanisms; it is not difficult to envisage that a similar approach could be adopted here.

Figure 8.5 shows the results of a test on a model slab with clamped edges (providing complete restraint of the edges against deformation) with a UDL. This is the slab equivalent of a fixed end beam, and by analogy with such a beam, moments of opposite sign would be expected at the centre and around the edges

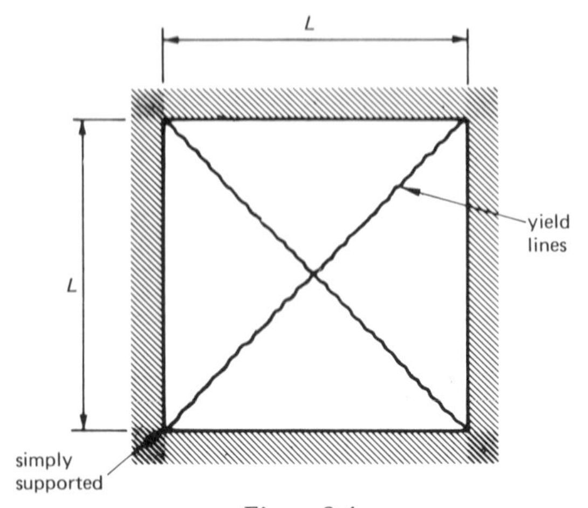

Figure 8.4

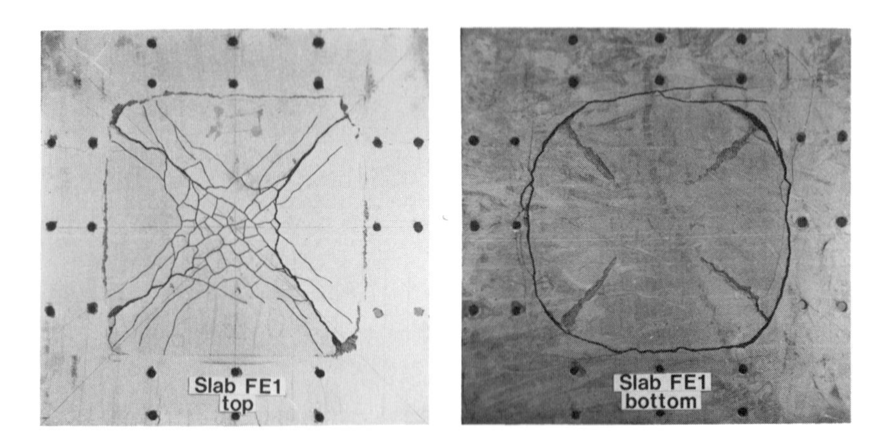

Figure 8.5 Test on square fixed-edge slab

of the slab. This is confirmed by figure 8.5. There are cracks on the top of the slab forming a very similar pattern to the simply supported slab in figure 8.1, including very large diagonal cracks. Around the edges there is evidence of crushing of the concrete. The bottom of the slab shows crushing along the diagonals and large cracks around the edge. The idealised collapse mechanism is shown in figure 8.6. Notice that the diagonal *positive* yield lines are represented differently from the *negative* yield lines around the edges.

8.2.2 A Notation for Yield Line Analysis

Diagrams of slab problems can become complicated because of the amount of

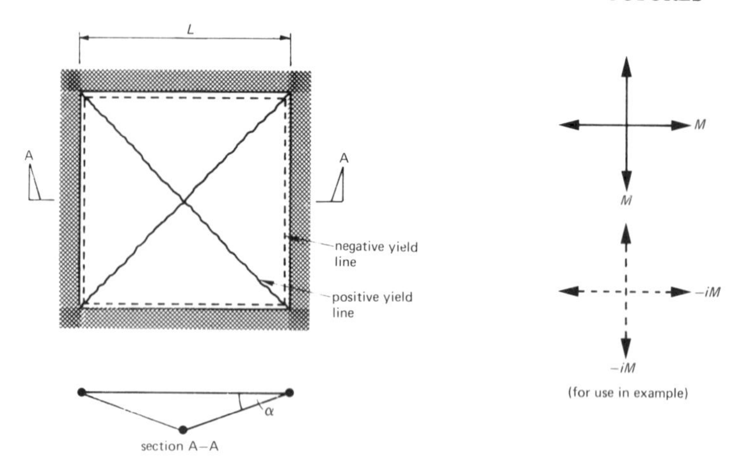

Figure 8.6

information to be presented. A useful notation has developed to simplify matters. It has been used to some extent already but is given here in more detail.

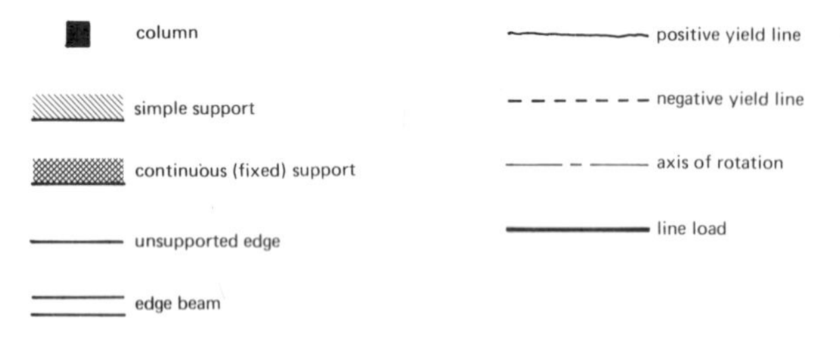

8.2.3 Theoretical Considerations of Yield Line Analysis

The major problem of yield line analysis is deciding on the collapse mechanism, although the calculations may well appear far worse to the beginner. Fortunately, many standard cases are well documented. [27] In general, any pattern of yield lines has to be postulated from experience and a useful set of rules (see section 8.2.4). There is no guarantee that the chosen pattern is the correct one: it will satisfy the mechanism and equilibrium conditions, but not necessarily the yield condition. Unfortunately, it is difficult, if not impossible, to check this, which means that in general any yield line solution will be an upper bound. Theoretically, yield line analysis is unsafe because an upper bound is an over-estimate of the strength of the slab. Physically it is safe because the analysis ignores two important factors.

(1) Moments of resistance of the slab are calculated ignoring strain hardening in the reinforcement. Figure 7.5 shows that at the low percentages of reinforcement in a slab, strain hardening significantly increases the moment, and thus the strength of the slab.

(2) Yield line theory is very much an idealisation of slab behaviour. It assumes that vertical loads are carried only by bending action. Tests show that this is not so. If there was only bending action the load deflection curve would have a continuously decreasing slope to failure. However, the central load deflection curve in figure 8.2 shows a decrease at first, but the slope then increases again, before decreasing finally to failure. The increase in slope is due to *membrane action*. The load is in fact partly carried by bending and partly by forces within the plane of the slab (membrane action). The more the slab deflects the more significant is the membrane action.

There are two distinct types of membrane action. Where the edges provide little or no restraint to horizontal movement (as is often the case with simply supported edges), *tensile* membrane action occurs. The system of inplane forces is indicated in figure 8.7.

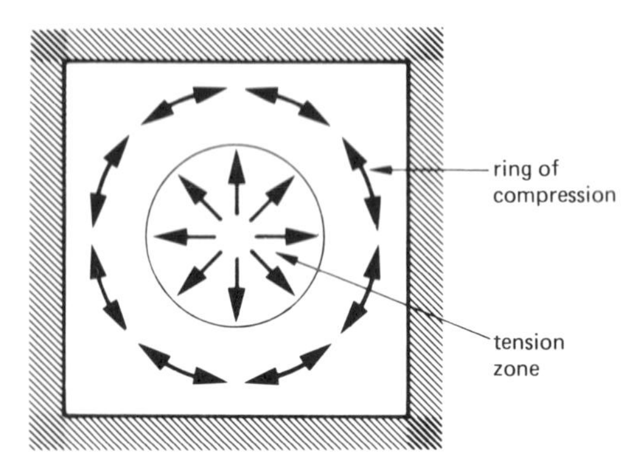

Figure 8.7

Tensile and compressive forces are required for horizontal equilibrium. The effect of tensile membrane action is to reduce the load causing bending in the tension zone and to increase it in the compression area. This preserves vertical equilibrium but reduces the maximum BM. Figure 8.8 illustrates the effect on a simply supported beam. This type of membrane action can increase the collapse load up to about 30 per cent above the yield line load.

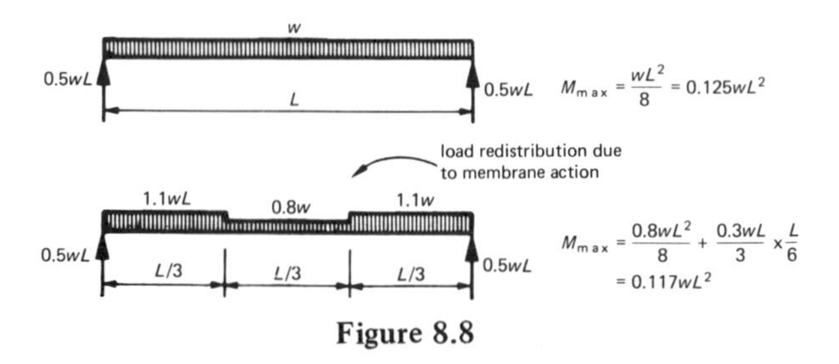

Figure 8.8

Compressive membrane action occurs in slabs with considerable horizontal edge restraint. As the cracks open the slab jams itself between the edges, as in figure 8.9. This jamming induces very large compressive forces in the slab so that it acts more like a shallow arch or dome. The result is an increase in load capacity of as much as 200 per cent above the yield line load.

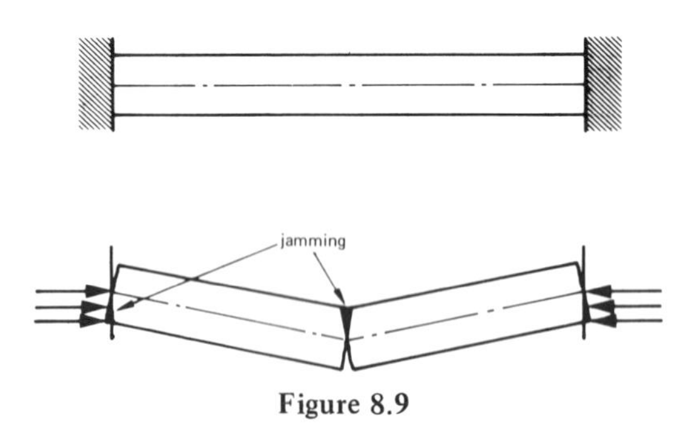

Figure 8.9

Ockleston [28] who was able to monitor the demolition of an old hospital, observed that it was virtually impossible to collapse a slab surrounded by other slabs (providing the horizontal restraint) by vertical loading. It is not possible to utilise the benefits of membrane action because it can be inconsistent and only occurs at large deflections. However, it does ensure the safety of yield line theory.

The formation of yield lines is predicted by the normal yield criterion. This states that a yield line will form at right-angles (normal) to a BM which has reached the moment of resistance of the slab. This can be visualised better by thinking in terms of the moment opening the concrete crack which coincides with the yield line. The moments at any point in a slab can be defined by a Mohr's circle and the yield line will form normal to one of the principal moments. [29] The criterion also allows a second yield line to form, at any point in the slab, normal to the first yield line. The criterion allows only two yield lines at any point.

8.2.4 Some Rules to Help Determine the Collapse Mechanism

The five rules below will help in finding collapse mechanisms. Use them to justify the mechanisms in figure 8.10.

(1) Yield lines are (usually) straight and are axes of rotation.

(2) Yield lines must end at a slab boundary.

(3) Axes of rotation lie along supported edges, cut unsupported edges, and pass over columns.

(4) The axes of rotation of adjacent rigid regions have a point of intersection (which may be at infinity).

(5) There are often negative yield lines along at least part of a fixed edge.

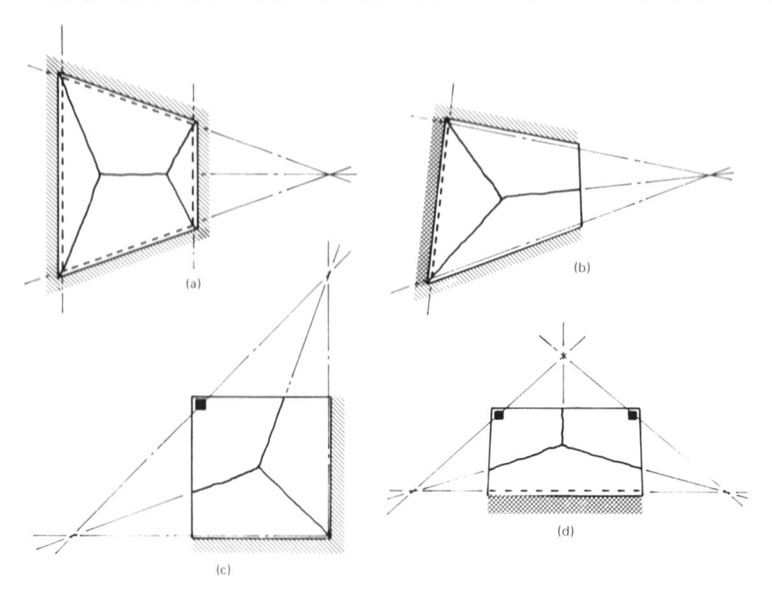

Figure 8.10

8.2.5 Moment Across a Yield Line

In order to carry out the calculations of yield line analysis it is necessary to know the moment across a yield line. Consider initially a slab with reinforcement in one direction only. In general a yield line will be inclined to the 'reinforcement', as in figure 8.11.

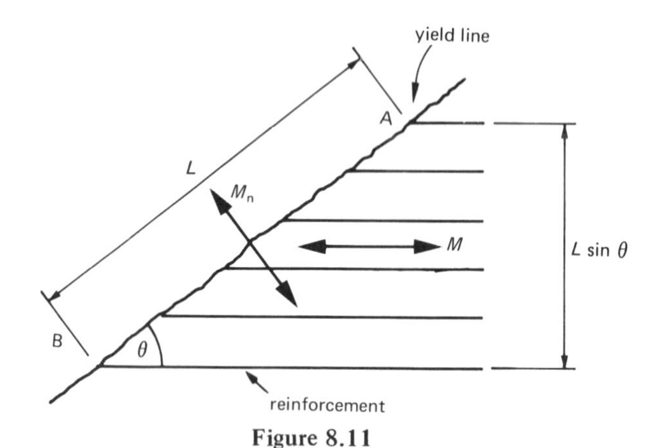

Figure 8.11

It is assumed that the reinforcement is yielding but remains straight, and that the reinforcement allows a moment of resistance, M, per unit width in the direction of the reinforcement. Along AB the total moment in the direction of the reinforcement is $ML \sin\theta$. Resolving this moment normal to the yield line gives

$$M_n L = ML \sin\theta \times \sin\theta$$

$$M_n = M \sin^2\theta \qquad (8.1)$$

where M_n is the moment per unit width across the yield line.

Of course, there is usually reinforcement in two directions and the resulting moment across the yield line can be found from equation 8,1, Figure 8.12 shows the most common situation of perpendicular reinforcement. (Notice the shorthand in the figure to identify the moment of resistance of each set of reinforcement. The arrows indicate the direction of the reinforcement.) Using equation 8.1

$$M_n = M \sin^2\theta + \mu M \sin^2(90 + \theta)$$

$$M_n = M \sin^2\theta + \mu M \cos^2\theta \qquad (8.2)$$

The factor μ is used to show that there can be different moments of resistance in the two reinforcement directions. If $\mu \neq 1$ the reinforcement is called *orthotropic.* The special case when $\mu = 1$ is called *isotropic* reinforcement.

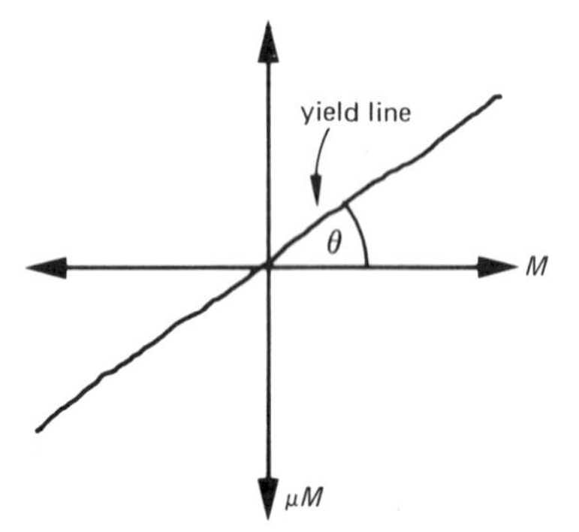

Figure 8.12

When $\mu = 1$

$$M_n = M(\sin^2\theta + \cos^2\theta)$$

$$M_n = M \qquad (8.3)$$

so that the moment across the yield line is the same whatever the inclination of the yield line.

It is worth mentioning the less common case of *skew* reinforcement, where the reinforcement directions are not perpendicular. A typical arrangement is shown in figure 8.13. Again using equation 8.1

$$M_n = M \sin^2\theta + \mu M \sin^2(\theta - \alpha) \qquad (8.4)$$

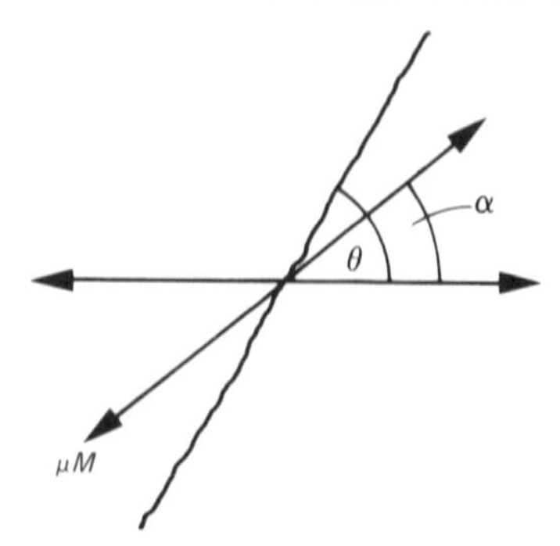

Figure 8.13

There are two further points to note about equations 8.1 to 8.4.

(1) Remember that M_n, M and μM are all moments *per unit width.* The total moment across a yield line is M_n multiplied by the length of the yield line.

(2) M and μM depend on the amount of reinforcement in the slab. They can be found by analysing a beam section of unit width, as described in chapter 7. Of course, any of the methods given in CP 110 can be used (including the formulae) when designing to that code. When the slab is being designed, M and μM would be found from the yield line analysis and it would then be necessary to provide enough reinforcement to develop those moments of resistance.

8.2.6 The Calculations

Once the collapse mechanism is determined, the slab can be analysed. The method will be illustrated by a series of examples which, as usual, have been graded to introduce various concepts or difficulties. The calculations are based on the virtual work method.

8.2.6.1 Square, Simply Supported Slab with UDL (q per unit area)

The distribution of yield lines for such a slab was discussed in section 8.2.1 and is shown again in figure 8.14a. The diagram also indicates that the slab has isotropic reinforcement of M per unit width.

The displacement of the mechanism is fully described by a unit (virtual) vertical displacement at the centre of the slab, remembering that the triangles between the yield lines remain flat. To find the external work

$$\text{total load on triangle ABE} = q \times \left(\frac{1}{2} \times \mathrm{L} \times \frac{L}{2}\right) = \frac{qL^2}{4}$$

$$= \text{UDL} \times \text{surface area of ABE}$$

the centroid of this load is at G, such that FG : FE = 1 : 3, from the basic

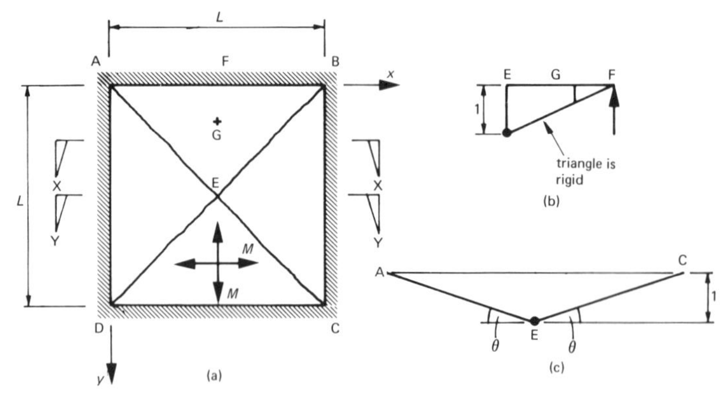

Figure 8.14

properties of a triangle. If E moves downwards by 1 (unit distance), G moves $\frac{1}{3}$, from the similar triangles in figure 8.14b

$$\text{external work by load on ABE} = \frac{1}{3} \times q\frac{L^2}{4} = \frac{qL^2}{12}$$

$$= \underset{\text{by centroid}}{\text{distance moved}} \times \text{load on ABG}$$

since the four triangles are identical

$$\text{total external work} = 4 \times \frac{qL^2}{12} = \frac{qL^3}{3} \tag{8.5}$$

The calculations for the external work are almost identical to those used for the steel frames. In its most general form the external work can be written

$$\text{external work} = \Sigma\left[\int_A q\Delta \, \mathrm{d}A\right] \tag{8.6}$$

all rigid regions — integral of load on an element of area multiplied by distance through which the element of the area moves.

The integral is necessary because the load need not be constant over the rigid region.

The internal work is the work absorbed by rotation of the yield lines. Figure 8.14c shows a cross-section through the slab along the diagonal AC. Remembering that the rigid regions remain flat, the rotation of the yield line BED is constant along its whole length and is equal to 2θ as shown in the figure.

$$\text{internal work for yield line BED} = M_n \times \sqrt{(2)}L \times 2\theta$$

since the slab is isotropically reinforced, $M_n = M$. From figure 8.14c

$$\theta = \frac{1}{EC} = \frac{1}{\left(\frac{\sqrt{(2)L}}{2}\right)} = \frac{\sqrt{2}}{L}$$

$$\text{internal work for BED} = M \times \sqrt{(2)}L \times 2 \times \frac{\sqrt{2}}{L} = 4M$$

because of the symmetry of the slab, internal work for AEC is identical and

$$\text{total internal work} = 2 \times 4M = 8M \tag{8.7}$$

Notice that the dimensions of the slab cancel out from the internal work expression. Again the internal work can be expressed generally

$$\text{internal work} = \Sigma\left[\theta \int_s M_n \, ds\right] \tag{8.8}$$

every yield line

total rotation of yield line

total moment across the yield line

The integral is necessary because the moment across the yield line can vary along the length, due to a change in thickness of the slab or in reinforcement spacing.

Imposing equilibrium by equating external and internal work gives the solution

$$\frac{qL^2}{3} = 8M$$

so that

$$q = \frac{24M}{L^2} \text{ or } M = \frac{qL^2}{24} \tag{8.9}$$

There are two forms of the solution depending on whether the problem is one of analysis (M and L given) or design (q and L given). The general expression for the work equation is

$$\Sigma\left[\int_A q\Delta \, dA\right] = \Sigma\left[\theta \int_s M_n \, ds\right] \tag{8.10}$$

all rigid rigions

every yield line

In this example the exact position of the yield lines was known and the reinforcement was isotropic. Consequently there was no difficulty in finding the moment across the yield line. Most problems are not so helpful. Look at any of the patterns in figure 8.8; in every case the patterns are complicated and not explicitly defined. Throw in orthotropic reinforcement and it becomes practically impossible to find the moment across each yield line. Fortunately the internal work can be found by another method.

Consider the yield line AE. The reinforcement in the x-direction is yielding and the moment of resistance is M per unit width

$$\text{total moment in } x\text{-direction crossing AE} = \frac{ML}{2}$$

The sections marked X–X and Y–Y in figure 8.14 are shown in figure 8.15. The sections show that because of the rigid regions the plastic rotation of this moment is θ_x along the whole length of AE. From the geometry in the figure

$$\theta_x = 1 \bigg/ \frac{L}{2} = \frac{2}{L}$$

Internal work done by the moment in the x-direction is

$$\text{internal work} = \frac{ML}{2} \times \frac{2}{\text{L}} = M$$

In the y-direction, total moment crossing AE = $ML/2$. Similar sections to X–X and Y–Y could be drawn to show that

$$\theta_y = \frac{2}{L}$$

so that the internal work is again M.

$$\text{total internal work for AE} = M + M = 2M$$

The other three parts of the yield line pattern are identical to AE, which means

$$\text{total internal work} = 4 \times 2M = 8M \text{ (as before)} \tag{8.11}$$

In this alternative method the moment across the yield line and the plastic rotation of the yield line are divided into vector components.

$$\overrightarrow{\int_s M_x \, \mathrm{d}s} = \overrightarrow{\int M_x \, \mathrm{d}y} + \overrightarrow{\int M_y \, \mathrm{d}x}$$

$$\vec{\theta} = \overrightarrow{\theta_x} + \overrightarrow{\theta_y}$$

$$\vec{\theta} \cdot \vec{M_\text{n}} = (\overrightarrow{\theta_x + \theta_y}) \cdot \left(\overrightarrow{\int M_x \, \mathrm{d}y} + \overrightarrow{\int M_y \, \mathrm{d}x} \right)$$

$$= \vec{\theta_x} \cdot \overrightarrow{\int M_x \, \mathrm{d}y} + \vec{\theta_y} \cdot \overrightarrow{\int M_y \, \mathrm{d}x} \tag{8.12}$$

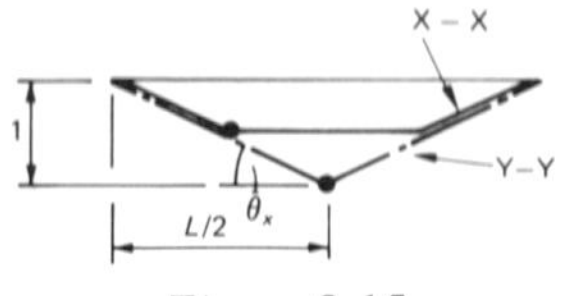

Figure 8.15

If this is summed over every yield line the general expression for internal work can be expressed in two ways

$$\text{internal work} = \Sigma\left[\theta \int_s M_x \,\mathrm{d}s\right] = \Sigma\left[\theta_x \int M_x \,\mathrm{d}y + \theta_y \int M_y \,\mathrm{d}x\right] \tag{8.13}$$

There is no advantage in using this method for this example, but as will be seen in some of the later examples it is often the only possible method.

8.2.6.2 Square Fixed Edge Slab with UDL

This slab is shown in figure 8.6 (p. 156). There are two sets of moments of resistance. Those shown by the full arrows are provided by the reinforcement in the bottom of the slab (resisting sagging moments), the broken arrows show the moments of resistance provided by the top reinforcement. In both cases the reinforcement is isotropic.

External work. The negative yield lines around the edge of the slab are the only difference from the previous example If the centre of the slab is given a unit vertical deflection, the external work is identical to the previous example

$$\text{external work} = \frac{qL^2}{3}$$

Internal work. The positive yield lines are the same as in the previous example

$$\text{internal work at positive yield lines} = 8M$$

$$\text{internal work at negative yield lines} = 4 \times iML \times \frac{1}{L/2}$$

no. of YLs (↗ 4); total moment across YL (↗ iML); rotation of YL (↑ $\frac{1}{L/2}$)

$$= 8iM$$

Notice that the sign of the moment and rotation can be ignored, the internal work always comes out positive. The rotation of the negative yield line is the angle α in the cross-section shown in figure 8.5

$$\text{total internal work} = 8(1 + i)M$$

For equilibrium

$$\frac{qL^2}{3} = 8(1 + i)M$$

At the corners the mechanism requires one positive and two negative yield lines to meet at one point. As was explained earlier, the moment across any point on a yield line must be one of the principal moments at that point. The yield criterion allows both principal moments to reach the corresponding moment of resistance so that there can be two yield lines, perpendicular to each other, at any point. At the corners there are apparently three yield lines, which is not allowable. The calculations for the mechanism are perfectly valid, but the answer is obviously an upper bound.

8.2.6.3 *Rectangular, Orthotropic Slab with Fixed Edges*

This example is derived from the model rectangular slab shown in figure 8.16.

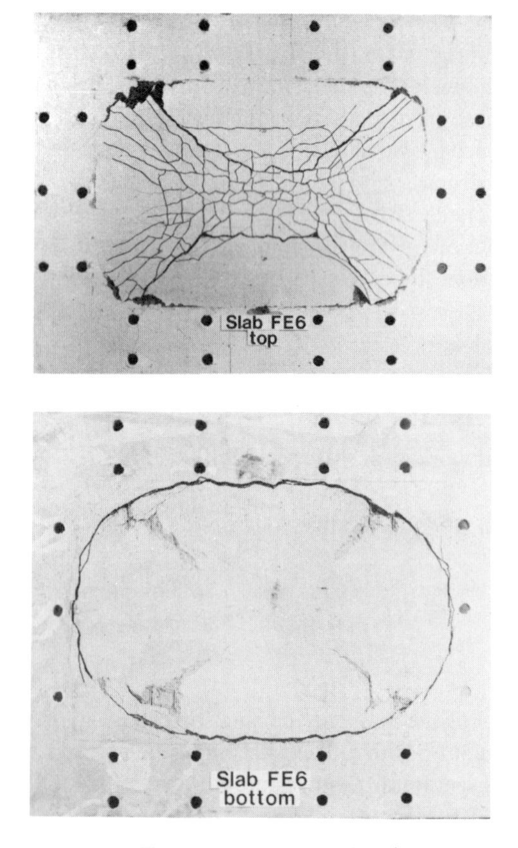

Figure 8.16 Test on rectangular fixed-edge slab

The large cracks and crushed concrete give the yield line pattern in figure 8.17. Tests show that the point of intersection of the positive yield lines, defined by the dimension βL, depends on the ratio of the side lengths and the factor μ which defines the orthotropic reinforcement. Consequently the actual position of the inclined yield lines is unknown. This means that some complicated expressions would be needed to define the moment across and the rotation of these yield lines. The vector method gives the internal work without too much trouble. Once the work equation has been found, it is necessary to find the critical value of β.

The loading on the slab is assumed to be: UDL q per unit area, line load Q per unit length along the long centre line. The deformation of the mechanism is defined by the unit deflections shown in figure 8.17

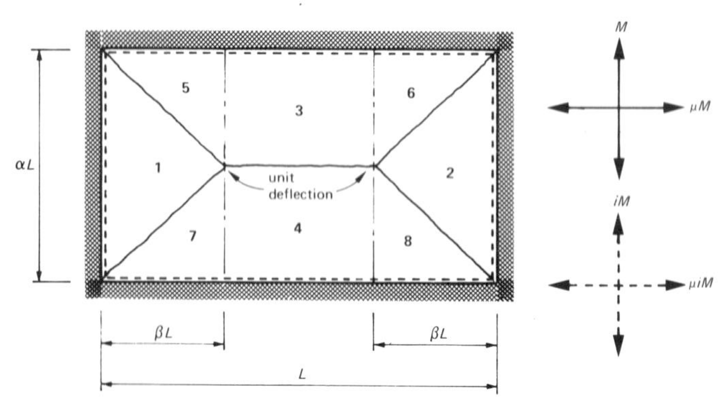

Figure 8.17

$$\text{external work} = q\underset{\mathbf{1,2}}{\frac{\alpha L \times \beta L}{2} \times \frac{1}{3} \times 2} + q\underset{\mathbf{3,4}}{\frac{\alpha L \times (1-2\beta)L}{2} \times \frac{1}{2} \times 2}$$

$$+ q\underset{\mathbf{5,6,7,8}}{\frac{\alpha L}{2} \times \frac{\beta L}{2} \times \frac{1}{3} \times 4} + \underset{\mathbf{1,2}}{Q\,\beta L \times \frac{1}{2} \times 2} + \underset{\mathbf{3,4}}{Q \times (1-2\beta)L \times 1}$$

The rigid regions have been sub-divided into triangles and rectangles to simplify the calculations. To keep track of the calculations it is worth numbering each of these. The bold numbers under each expression correspond to the numbers in figure 8.17. In this way none of the loading is omitted from the external work.

$$\text{external work} = q\frac{\alpha\beta L^2}{3} + q\left(\frac{\alpha}{2} - \alpha\beta\right)L^2 + q\frac{\alpha\beta L^2}{3}$$

$$+ Q\beta L + Q(1-2\beta)L$$

$$= q\left(\frac{\alpha}{2} - \frac{\alpha\beta}{3}\right)L^2 + Q(1-\beta)L$$

The internal work needs some thought. Consider the yield lines shown in figure 8.18. Using the vector components in the x-direction only

$$\text{internal work} = \underset{\substack{\text{positive}\\ \text{yield lines}}}{\mu M \times \alpha L \times \theta_x} + \underset{\substack{\text{negative}\\ \text{yield lines}}}{\mu i M \times \alpha L \times \theta_x}$$

$$= \mu M(1 + i) \times \alpha L \times \frac{1}{\beta L}$$

$$= \mu M(1 + i) \frac{\alpha}{\beta}$$

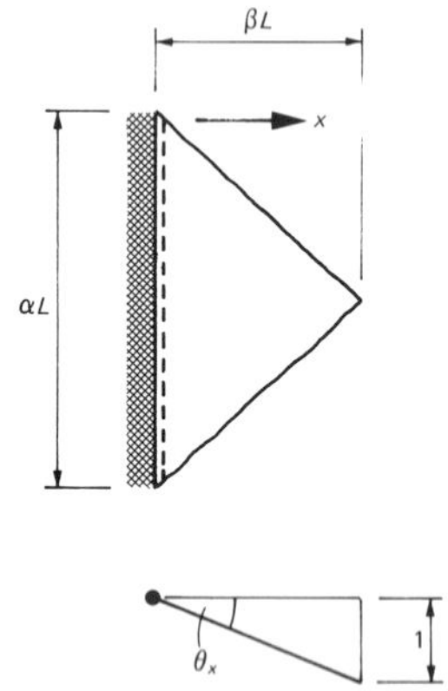

Figure 8.18

The same expression could be used for the other end of the slab. All the other yield lines are parallel to the x-direction and have no vector component in that direction. This means that the total internal work due to x-direction components

$$= \mu M (1 + i) \frac{\alpha}{\beta}$$

Consider now the part of the slab shown in figure 8.19. The trapezoidal rigid

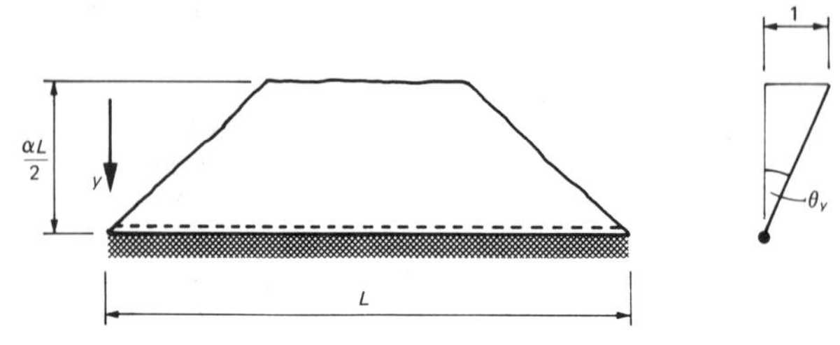

Figure 8.19

region is flat so that the angle θ_y is the plastic rotation for the y-components of every yield line in figure 8.19

$$\text{internal work} = \underset{\substack{\text{positive}\\ \text{YLs}}}{ML\theta_y} + \underset{\substack{\text{negative}\\ \text{YLs}}}{iML\theta_y}$$

$$= M(1+i)L \times \frac{1}{\alpha L/2}$$

$$= 2M(1+i)\frac{1}{\alpha}$$

hence

$$\text{total internal work} = 2\mu M(1+i)\frac{\alpha}{\beta} + 4M(1+i)\frac{1}{\alpha}$$

$$= 2M(1+i)\left(\frac{2}{\alpha} + \frac{\mu\alpha}{\beta}\right)$$

and the work equation is

$$2M(1+i)\left(\frac{2}{\alpha} + \frac{\mu\alpha}{\beta}\right) = q\left(\frac{\alpha}{2} - \frac{\alpha\beta}{3}\right)L^2 + Q(1-\beta)L \tag{8.14}$$

Equation 8.14 relates the moment of resistance M to the loadings q and Q. The critical value of the variable β is the value which gives the largest value of M for given q and Q or vice versa. This maximising process can be achieved by finding the value of β which satisfies

$$\frac{dM}{d\beta} = 0 \quad \text{or} \quad \frac{dq}{d\beta} = 0 \tag{8.15}$$

(The second equation assumes that there is some relationship between q and Q.) To illustrate what happens assume that $Q = 0$ to simplify the mathematics. Now

$$2M(1+i) = q\left(\frac{\alpha}{2} - \frac{\alpha\beta}{3}\right)L^2 \Big/ \left(\frac{2}{\alpha} + \frac{\mu\alpha}{\beta}\right) \tag{8.16}$$

and using the quotient rule for differentiation

$$\frac{2(1+i)}{qL^2}\frac{dM}{d\beta} = \frac{\left(\frac{2}{\alpha} + \frac{\mu\alpha}{\beta}\right)\left(-\frac{\alpha}{3}\right) - \left(-\frac{\mu\alpha}{\beta^2}\right)\left(\frac{\alpha}{2} - \frac{\alpha\beta}{3}\right)}{\left(\frac{2}{\alpha} + \frac{\mu\alpha}{\beta}\right)^2} = 0$$

which can be simplified to

$$\left(\frac{2}{\alpha}+\frac{\mu\alpha}{\beta}\right)\left(-\frac{\alpha}{3}\right)+\left(\frac{\mu\alpha}{\beta^2}\right)\left(\frac{\alpha}{2}-\frac{\alpha\beta}{3}\right)=0$$

$$-\frac{2}{3}-\frac{\mu\alpha^2}{3\beta}+\frac{\mu\alpha^2}{2\beta^2}-\frac{\mu\alpha^2}{3\beta}=0$$

$$\frac{2}{3}+\frac{2\mu\alpha^2}{3\beta}-\frac{\mu\alpha^2}{2\beta^2}=0$$

multiplying through by $6\beta^2$ gives the quadratic equation

$$4\beta^2+4\mu\alpha^2\beta-3\mu\alpha^2=0$$

which can be solved to give

$$\beta=\frac{-4\mu\alpha^2\pm\sqrt{(16\mu^2\alpha^4+48\mu\alpha^2)}}{8}$$

$$=\frac{-\mu\alpha^2}{2}\pm\frac{1}{2}\sqrt{(\mu^2\alpha^4+4\mu\alpha^2)}$$

The only possible value of β is the positive root (a negative value would indicate yield lines outside the slab) and thus

$$\beta_{\text{crit}}=\frac{-\mu\alpha^2}{2}+\frac{1}{2}\sqrt{(\mu^2\alpha^4+4\mu\alpha^2)} \qquad (8.17)$$

This can now be substituted back into 8.16 to give the critical solution of the work equation. Equation 8.17 confirms that β_{crit} depends on the factors μ and α which define the orthogonal reinforcement and the ratio of the side lengths of the slab.

The reader might like to verify that when $Q = kqL$ the quadratic equation for β_{crit} is

$$4\beta^2+4(\mu\alpha^2+3\mu\alpha k)-3(\mu\alpha^2+2\mu\alpha k)=0$$

8.2.6.4 Rectangular, Orthotropic Slab with Mixed Supports

The maximising process can become very complicated when several variables are involved in the mechanism. Consider the slab shown in figure 8.20. There are two variables β_1 and β_2 because of the different boundary conditions on the long edges. The derivation of the work equation is very similar to the previous example The reader ought to verify that it is

$$\frac{7M}{4}\left[\frac{1.5\,\beta_2\,(1-\beta_1)+\beta_2+0.3265\beta_1\,(1-\beta_1)}{\beta_1\beta_2\,(1-\beta_1)}\right]=\frac{14q}{3}(3-2\beta_2)$$

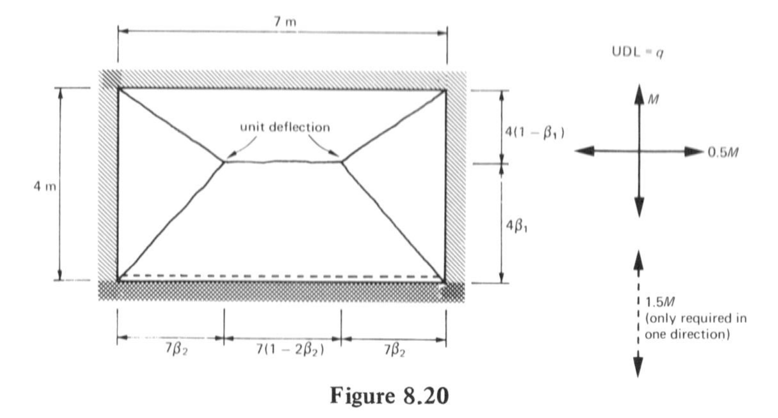

Figure 8.20

The solution of the simultaneous equations

$$\frac{\partial M}{\partial \beta_1} = 0 \qquad \frac{\partial M}{\partial \beta_2} = 0$$

will give the critical values of β_1 and β_2, but the differentiation is quite formidable. The alternative is to use a trial and error technique. Simply choose likely ranges for β_1 and β_2 and calculate the work equation for various combinations of β_1 and β_2. It may seem long-winded, but in fact is very quick provided a calculator is available. After a little experience it is possible to 'home in' on the critical values.

In this case β_1 is going to be about 0.5, so start with a range 0.3 to 0.7, and β_2 is less than 0.5, start with the range 0.15 to 0.35. The calculations are given in table 8.1

Table 8.1 Values of M/q

β_2 \ β_1	0.3	0.35	0.4	0.45	0.5	0.55	0.6	0.65	0.7
0.15	0.603	0.663	0.713	0.754	0.785	0.805	0.814	0.811	0.793
0.2			0.726	0.770	0.803	0.825	0.835	0.832	0.812
0.25					0.803	0.826	0.836	0.832	0.812
0.3						0.815	0.825	0.821	
0.35						0.796	0.807	0.803	

Compare the results from the table with the exact values which are shown in brackets

$$\beta_{1\,\text{crit}} = 0.6 \qquad (0.613)$$

$$\beta_{2\,\text{crit}} = 0.25 \qquad (0.227)$$

$$M = 0.836q \qquad (0.838q)$$

The difference is not significant.

8.2.7 Slabs with Edge Beams

It is very common that the RC slab is supported around the boundaries by edge beams which span between columns at the corners of the slab. Under load the slab and beams will all deflect. The boundary conditions are now different from any so far considered, and merit further examination.

Tests on beam and slab arrangements produce interesting results. Various modes of failure occur, depending on the relative moments of resistance of the beams and slab. Figure 8.21 illustrates several mechanisms which have resulted from slab tests. The trouble is that it is not obvious which mechanism will occur in any problem.

It would seem that the analyst must try every mechanism to find which one occurs at the lowest load. Consider the square slab in figure 8.22. The work equations for the various mechanisms are as follows.

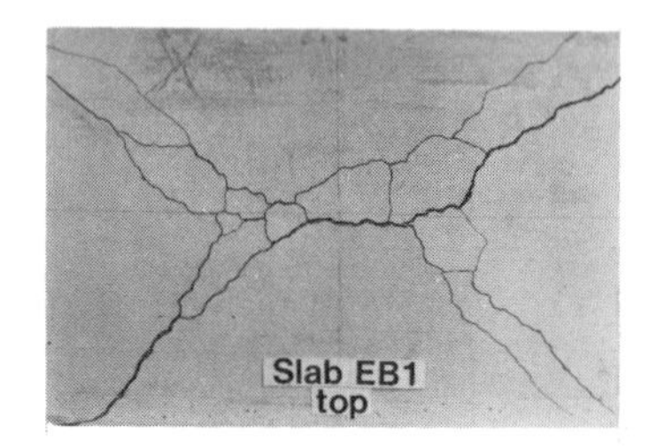

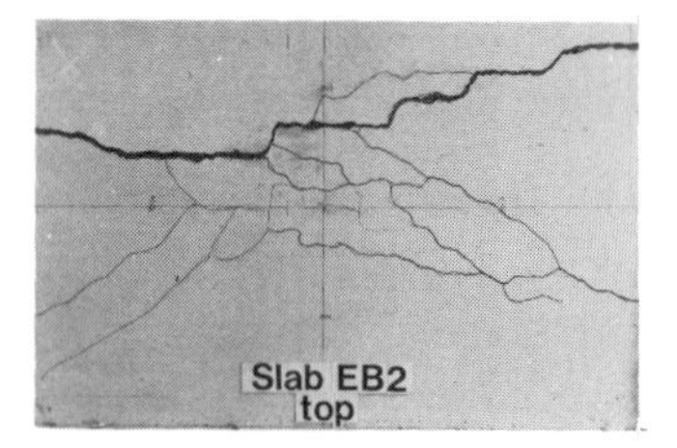

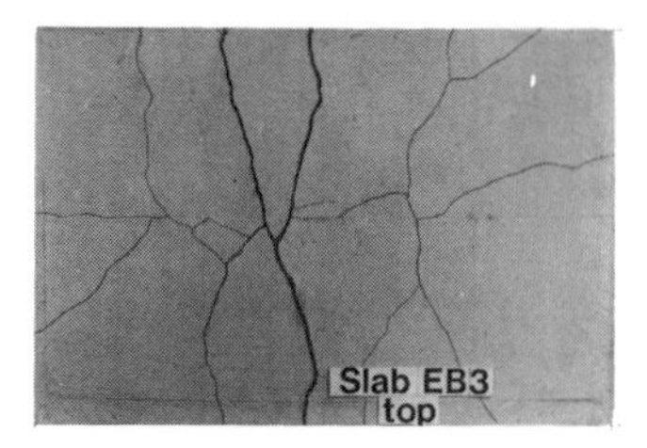

Figure 8.21 Tests on slabs with edge beams

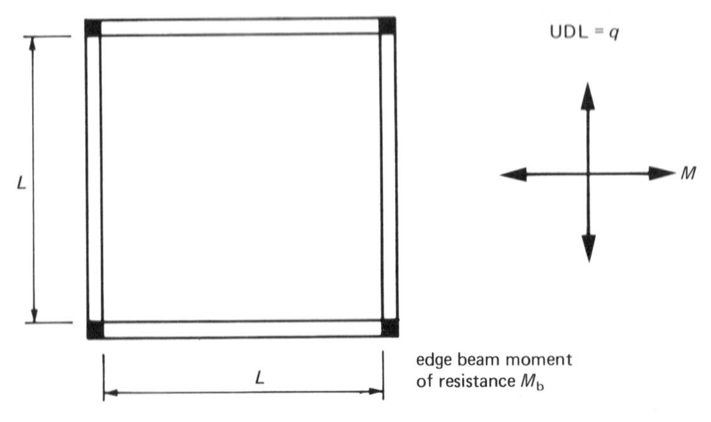

Figure 8.22

(a) The edge beams are so strong that the diagonal mechanism forms in the slab. Only positive yield lines are required because the beams can rotate after torsion failure in the corners. (Look at the left-hand corner of slab EB1 in figure 8.21.) The work equation for this mechanism has already been found (see section 8.2.6.1).

$$q = \frac{24M}{L^2} \tag{8.18}$$

(b) The work equation can be found by assuming unit deflection along the yield line. (See p. 174.)

$$\text{internal work} = ML\left(2 \times \frac{1}{L/2}\right) + 2M_b\left(2 \times \frac{1}{L/2}\right)$$

$$= 4M + \frac{8M_b}{L}$$

$$\text{external work} = qL \times \frac{L}{2} \times \frac{1}{2} \times 2$$

$$= \frac{qL}{2}$$

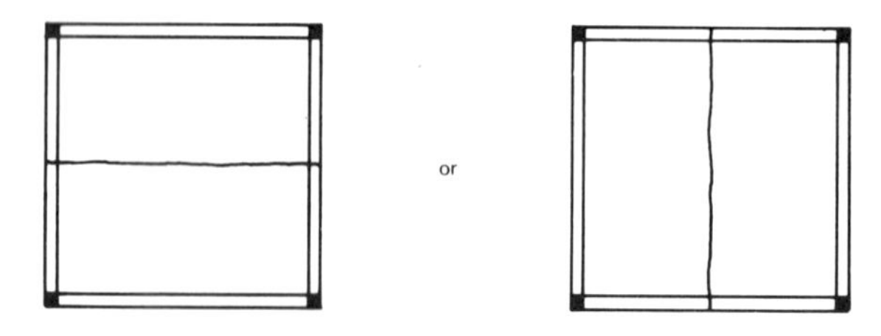

for collapse

$$q = \frac{8M}{L^2}\left(1 + \frac{2M_b}{ML}\right) \tag{8.19}$$

(c) Remembering the rules for the axes of rotation it is possible to determine the relative displacements in the mechanism.

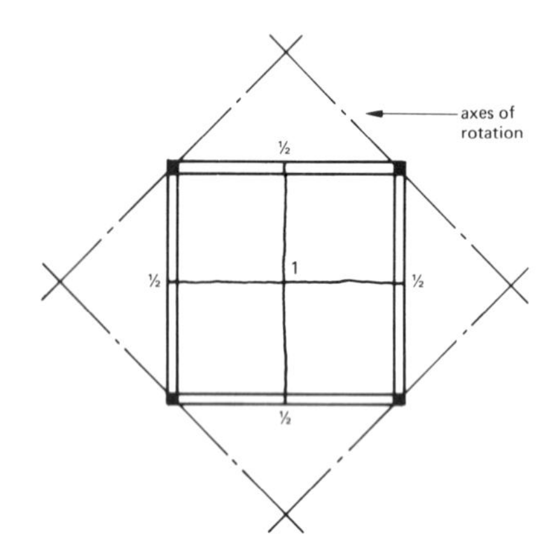

$$\text{internal work} = \left[ML\left(2 \times \frac{1/2}{L/2}\right) + 2M_b \left(2 \times \frac{1/2}{L/2}\right)\right] \times 2$$

$$= 4M + \frac{8M_b}{L}$$

$$\text{external work} = q \times \frac{L}{2} \times \frac{L}{2} \times \frac{1}{2} \times 4$$

$$= \frac{qL^2}{2}$$

thus the work equation gives

$$q = \frac{8M}{L^2}\left(1 + \frac{2M_b}{ML}\right) \tag{8.20}$$

which is identical to the previous mechanisms.

(d) This final mechanism can occur when one of the edge beams is weaker than the others.

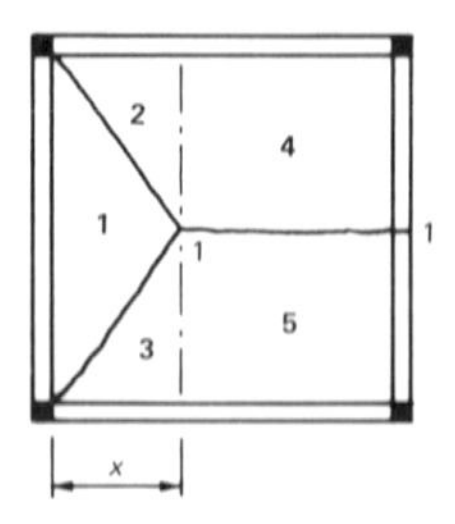

internal work (using vector components) $= ML\left(2 \times \frac{1}{L/2}\right) + M_b\left(2 \times \frac{1}{L/2}\right) + ML\frac{1}{x}$

$$= 4M + \frac{4M_b}{L} + \frac{ML}{x}$$

$$= 4M\left(1 + \frac{M_b}{ML} + \frac{L}{4x}\right)$$

external work $= q\frac{Lx}{2} \times \frac{1}{3} + q\frac{L}{2} \times \frac{x}{2} \times \frac{1}{3} \times 2 + q(L - x) \times \frac{L}{2} \times \frac{1}{2} \times 2$

1 **2, 3** **4, 5**

$$= q\frac{Lx}{6} + q\frac{Lx}{6} + q\frac{L^2}{2} - q\frac{Lx}{2}$$

$$= \frac{qL}{2}\left(L - \frac{x}{3}\right)$$

so that

$$q = \frac{8M}{L}\left(1 + \frac{M_b}{ML} + \frac{L}{4x}\right) \bigg/ \left(L - \frac{x}{3}\right) \tag{8.21}$$

It is necessary to find the critical value of the x using the maximising–minimising process.

It is interesting to consider the collapse loads given in equations 8.18 to 8.21. In equations 8.19 to 8.21 the collapse load depends on M_b/ML which is a measure of the relative magnitudes of the moments of resistance of the edge beams and slab. Figure 8.23 is a graph of collapse load against M_b/ML. The graph shows that when $M_b/ML > 1$ mechanism **(a)** is critical because the beams are sufficiently strong to resist collapse. When $M_b/ML = 1$, each mechanism has the same collapse load and is equally likely to occur.

The study of the interaction between beams and slab is a fascinating subject,

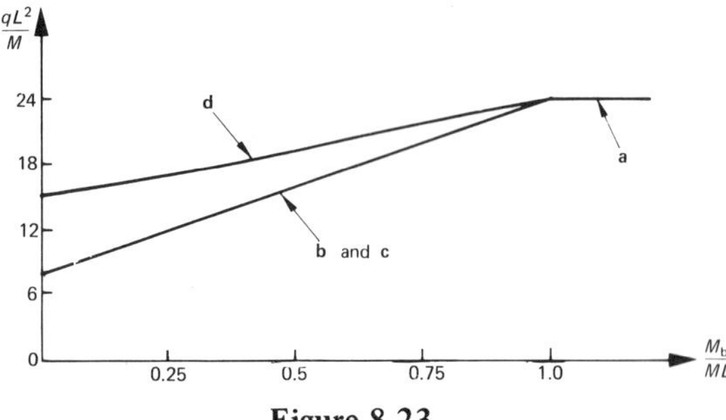

Figure 8.23

as Wood [30, 31] has shown. However, this section has also been included to illustrate a further point. In some slabs, particularly those with edge beams or unsupported edges, there can be several possible mechanisms, each with a different pattern of yield lines. It is necessary to analyse each one to find the collapse load of the slab. Figure 8.24 shows a typical example.

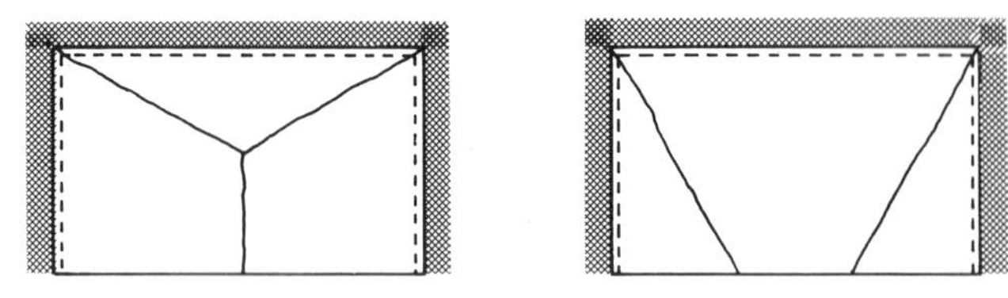

Figure 8.24

8.2.8 Yield Line Fans

8.2.8.1 *Slabs with Point Loads*

When a slab is tested with point loading, cracks form radially from the load. Flexural failure occurs when the radial cracks open out and a circumferential crack forms in the opposite face of the slab. This collapse mechanism is shown in figure 8.25. (There can be shear problems around point loads, but they will not

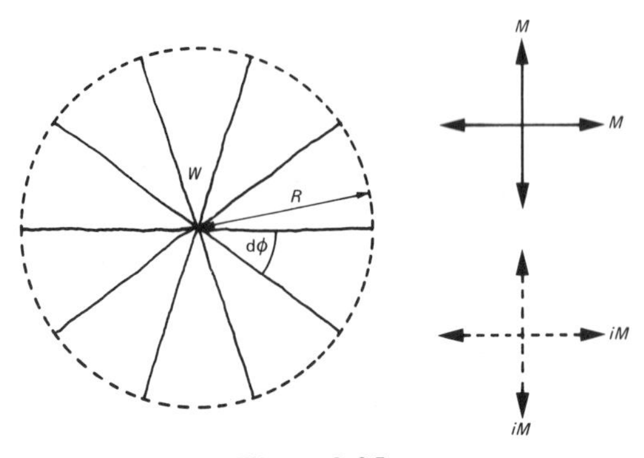

Figure 8.25

be considered in this book.) The mechanism is often called, for obvious reasons, a yield line fan. The value of W which causes collapse can be found in the usual manner by allowing W to move unit distance vertically.

$$\text{external work} = W \times 1$$
$$= W$$

It is not as simple to find the internal work. It is necessary to examine what is happening within the mechanism. The rigid region outside the negative yield line remains flat and horizontal. Each segment in the fan deforms as in figure 8.26. All deformation occurs by plastic rotation at the edges of the segment. It is necessary to consider vector components, but in this case the radial and tangential components are shown. As the reinforcement is isotropic the moments in both these directions are M and iM for positive and negative yield lines respectively.

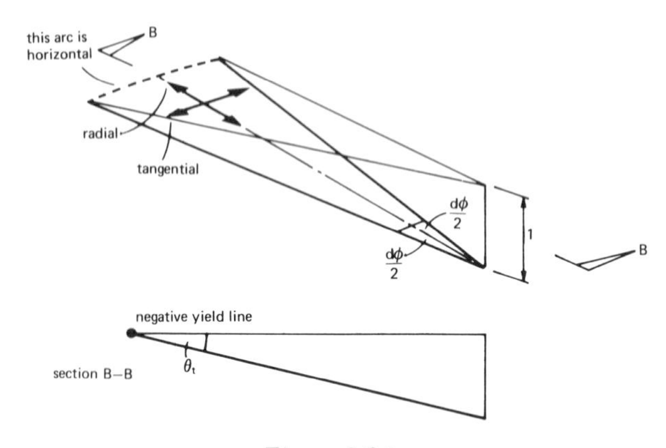

Figure 8.26

Since the region outside the mechanism is horizontal the arc of negative yield line is also horizontal. This means that there is no rotation about the radial direction within *the segment itself.* This means that for radial components

$$\text{internal work} = 0$$

For the tangential component

$$\text{internal work} = MR\mathrm{d}\phi \times \theta_t + iMR\mathrm{d}\phi \times \theta_t$$
$$= M(1 + i)R\mathrm{d}\phi \times \frac{1}{R}$$
$$= M(1 + i)\mathrm{d}\phi$$

The total internal work is the sum of the internal work for each segment

$$\text{total internal work} = M(1 + i)\phi$$

for a fan which subtends ϕ at the centre and

$$\text{total internal work} = M(1 + i)2\pi$$

for a complete fan.

The surprising thing about the internal work is that it is independent of the radius of the fan. This means that in a fixed edge slab *of any shape* provided there is top reinforcement everywhere in the slab, the collapse load must be

$$W_c = 2\pi M(1 + i) = 6.28M(1 + i)$$

since the complete fan can form within the slab because there is no restriction on the fan radius. The factor i is merely a device to express the moment across negative yield lines as a proportion of the positive moment. Of course, it can take any value, so that when $i = 0$, the result is for a slab with bottom reinforcement only. Thus a simply supported slab with a point load collapses when

$$W_c = 6.28M$$

For the designer with no knowledge of yield line fans the most likely choice of yield lines for a square fixed edge isotropic slab would be the diagonal pattern used in section 8.2.6.2. The diagonal pattern would give

$$W_c = 8M(1 + i)$$

This is 27 per cent higher than the collapse load from the fan!

8.2.8.2 *Another Use for Fans*

The example in section 8.2.6.2 was shown to be an obvious upper bound because there were too many yield lines in the corners. In the photograph of a square fixed edge slab (figure 8.5), it can be seen that the negative yield lines curve away from the corners and there are several cracks near the diagonals which could be interpreted as positive yield lines. Consequently the yield line pattern in figure 8.27 may well be a better idealisation of the collapse mechanism.

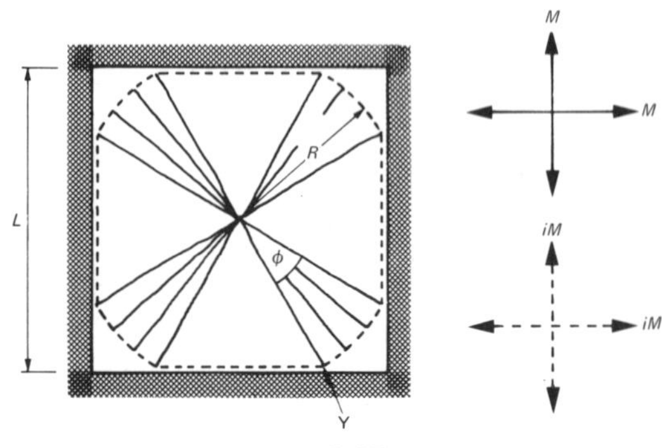

Figure 8.27

This mechanism contains four fans, each extending ϕ. Remembering that the centroid of a circular segment is two-thirds of the radius from the centre, the reader is left to show that the work equation for this mechanism is

$$q = \boxed{\frac{24M(1+i)}{L^2}} \left[\frac{\frac{\phi}{2} + \tan\left(\frac{\pi}{4} - \frac{\phi}{2}\right)}{\frac{\phi}{2}\ \sec^2\left(\frac{\pi}{4} - \frac{\phi}{2}\right) + \tan\left(\frac{\pi}{4} - \frac{\phi}{2}\right)} \right]$$

The first part of the equation (shown ringed) is identical to the diagonal yield line solution. The angle ϕ is a variable, and its critical value turns out to be almost 30°, with the corresponding solution

$$q = \frac{21.75M(1+i)}{L^2}$$

This is still only an upper bound. At point Y, although the negative yield line is now continuous, the intersection of the positive and negative yield lines is not at right-angles. Fox [32] has found the exact solution for the case where $i = 1$. His solution checked that the collapse mechanism satisfied the equilibrium, yield and mechanism conditions and involved some formidable mathematics and the use of a computer. The exact value of the collapse load is

$$q = \frac{42.851M}{L^2}$$

The overestimate of the collapse load is 12 per cent for the diagonal mechanism and 1.5 per cent for the fan mechanism. The extra labour of refining the mechanism has to be balanced against the improvement in accuracy, remembering the inherent variability of the concrete and the simplifying assumptions of yield line theory.

8.2.9 Design Details

The yield line method gives the moments of resistance required to resist collapse. At normal working loads the slab will still be in the elastic range. There would be considerable redistribution of moments as the collapse mechanism develops, and this would be accompanied by cracking of the concrete. To ensure that there are only small cracks at working load the reinforcement must be arranged, within practical limits, to suit the elastic moment distribution at working load.

In all the examples of fixed edge slabs the moment of resistance around the edges were chosen to be different to those in the centre of the slab, usually by means of the factor i. That factor must be set such that the moments of resistance are in about the same ratio as the corresponding elastic moments at normal working loads. CP 110 recommends that i should be in the range 1.0 to 1.5, although the results of elastic theory [29] indicate that the range should be 1.5 to 2.5.

8.2.10 Equilibrium Method

The virtual work approach can lead to complicated mathematical procedures and an alternative approach considering the equilibrium of each rigid region has been developed. This equilibrium method will not be described in this book, because the author feels that the virtual work method is easier for the beginner. The equilibrium method is extremely powerful, and has been described in detail in various text books. [1, 30, 33]

8.3 HILLERBORG'S STRIP METHOD

8.3.1 Background

This method was introduced in 1956 by Hillerborg. [34] Hillerborg's method manufactures a lower bound to the slab strength, and is thus inherently safe.

Before looking at the method it is necessary to compare the bending of a beam and a slab. This is done in figure 8.28, by examining the equilibrium of small beam and slab elements. In the beam element vertical equilibrium is ensured by shear forces at each end; in the slab element by shear forces on all four faces. To maintain moment equilibrium, moments must act at the ends of the beam element and around the slab element. The difference is that in the slab there are bending moments and twisting moments (reflecting the two-dimensional nature of the slab). The resulting equations give the requirement for bending equilibrium, and, since they say nothing about the nature of the material, hold up to collapse. In the beam the applied load is resisted by the bending moments, in the slab it is resisted by the bending moments about the x- and y-axes and also by the twisting moments. The equilibrium equation

$$\frac{\partial^2 M_x}{\partial_x{}^2} + \frac{\partial^2 M_y}{\partial y^2} - \frac{\partial^2 M_{xy}}{\partial x \partial y} + \frac{\partial^2 M_{yx}}{\partial x \partial y} = -q$$

is the starting point for Hillerborg's method. Assume that the twisting moments M_{xy} and M_{yx} are always zero, thus ignoring the twisting strength of the slab. (By deliberately ignoring the twisting strength, the overall strength must be underestimated.) This changes the equilibrium equation to

$$\frac{\partial^2 M_x}{\partial x^2} + \frac{\partial^2 M_y}{\partial y^2} = -q$$

Now assume that part of the load is carried by bending about the x-axis and

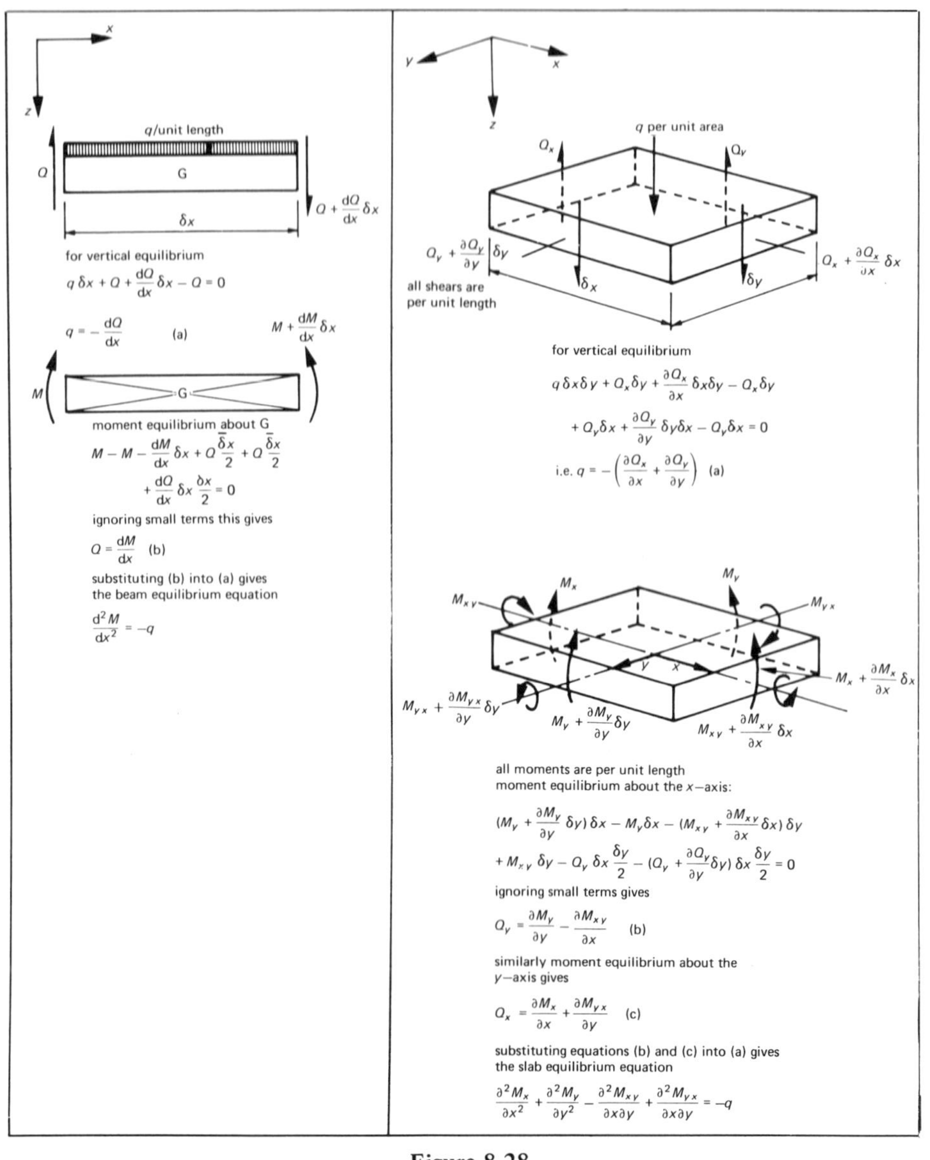

Figure 8.28

part by bending about the y-axis, so that

$$\frac{\partial^2 M_x}{\partial x^2} = -\alpha q$$

$$\frac{\partial^2 M_y}{\partial y^2} = -(1-\alpha)q$$

These are identical to the equation for beam bending. The slab problem has been converted to a beam bending problem. It is probably most convenient, but not essential, to assume that α is one or zero, so that

$$\frac{\partial^2 M_x}{\partial n^2} = -q \qquad \frac{\partial^2 M_y}{\partial y^2} = 0 \qquad \text{for } \alpha = 1$$

$$\frac{\partial^2 M_x}{\partial n^2} = 0 \qquad \frac{\partial^2 M_y}{\partial y^2} = -q \qquad \text{for } \alpha = 0$$

Thus it is assumed that at collapse the load is carried on beams spanning between opposite edges of the slab, as shown in figure 8.29. The lines of discontinuity merely divide the load on the slab surface. There is no critical position for each line, so there is no need for complicated mathematics. This is one of the advantages of the method. The problem now is to design the beams into which the slab has been converted. It is easiest to illustrate this by means of some examples.

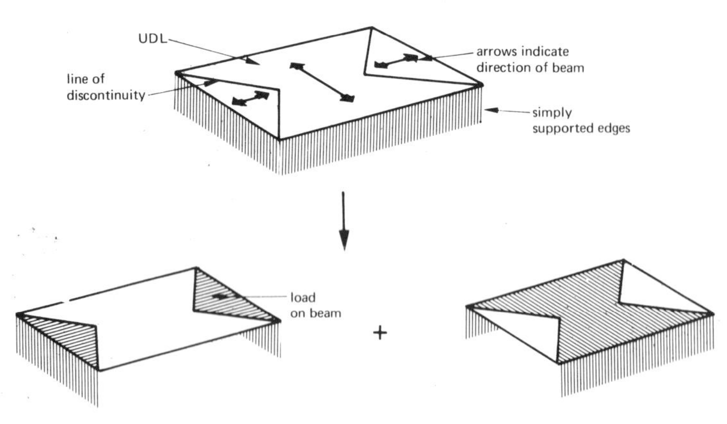

Figure 8.29

8.3.2 Simply Supported Rectangular Slab with UDL

Consider the slab in figure 8.30. It is fairly obvious that in the corners of the slab the load will be supported on both edges meeting at the corner. Hence it is reasonable to draw the line of discontinuity bisecting the angle in the corner, as shown. The arrows show the direction in which the load is carried, by bending, to the edges of the slab. In the middle of the long sides the load is only carried on a beam spanning between the long edges. The overall loading on the two beams is very similar to that shown in figure 8.29, but there is no need to have constant reinforcement in both beams. The beams have been sub-divided into strips by means of the broken lines. (It is these strips which give the method its full name.) The analysis of the individual strips is shown below.

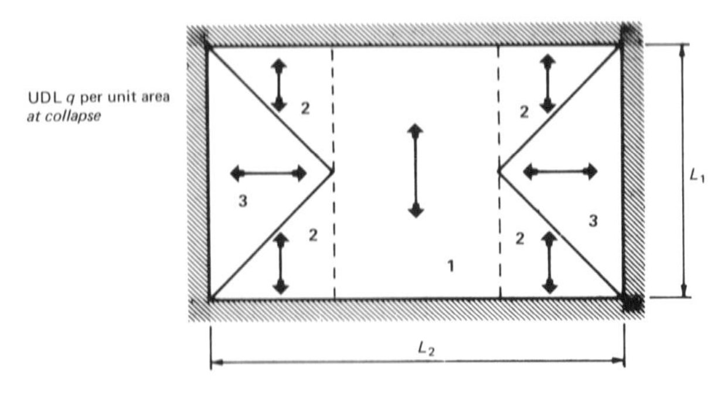

Figure 8.30

Strip **1** – consider a strip of unit width. The strip is a uniformly loaded beam with a span L_1

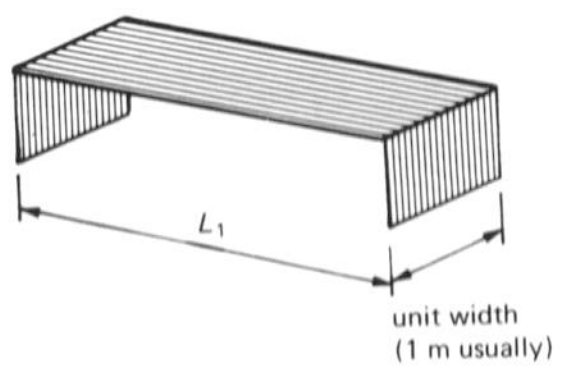

$$\text{Load on strip} = q \times 1$$

$$= q \text{ per unit length}$$

$$\text{maximum moment} = \frac{qL_1^2}{8} \text{ per unit width}$$

This is the design moment for the strip. The analysis of strips **2** and **3** is quite complicated. It can be greatly simplified by using loading of the same total intensity but evenly distributed across the width on the strip and then using a correction factor to give the true moment.

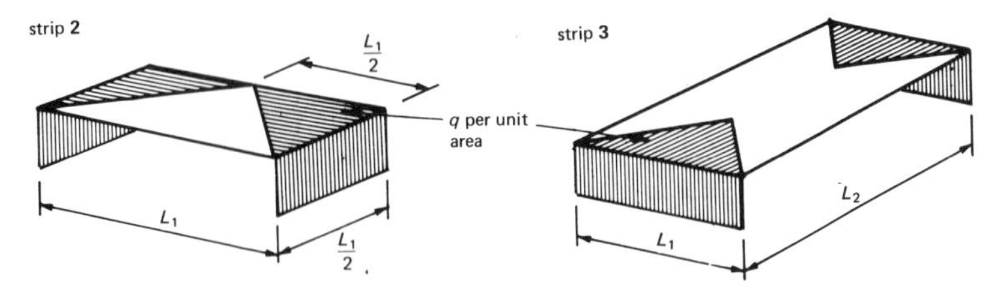

Strip **2** It is now possible to use a unit width of slab loaded as shown on p. 184

$$\text{maximum moment} = \frac{qL_1^2}{32} \text{ per unit width}$$

$$\text{true maximum moment} = \frac{KqL_1^2}{32} \text{ per unit width}$$

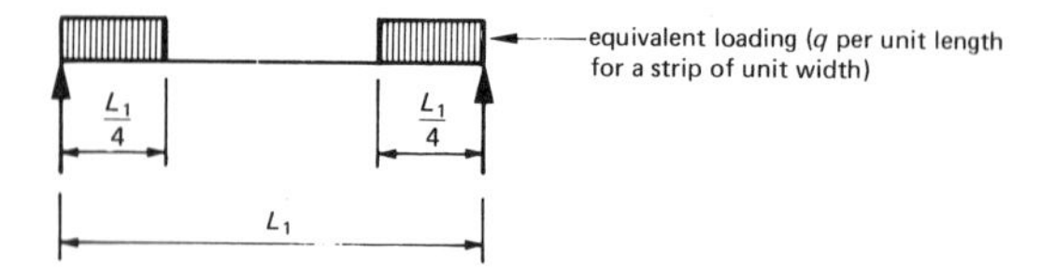

Strip 3

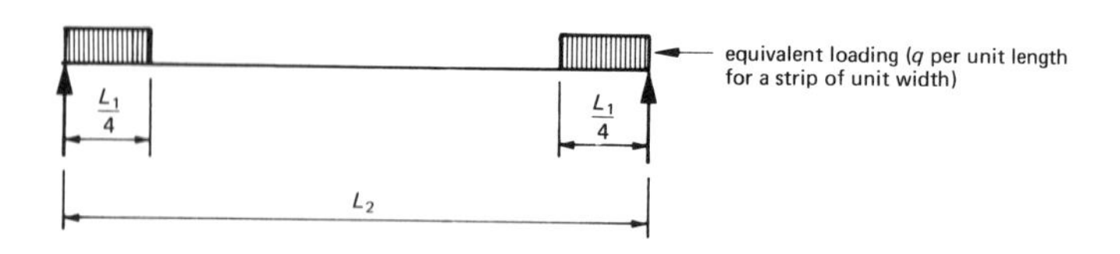

$$\text{true maximum moment} = \frac{K q L_1^2}{32}$$

The correction factor K is derived in the next section.

It is not essential to use the inclined line of discontinuity. It would be equally satisfactory to divide the load up as in figure 8.31. The only limitation on the strips is a purely practical one. There is a separate design moment and reinforcement requirement for each strip. It would be totally impractical to have too many strips because the reinforcement details would become too complicated.

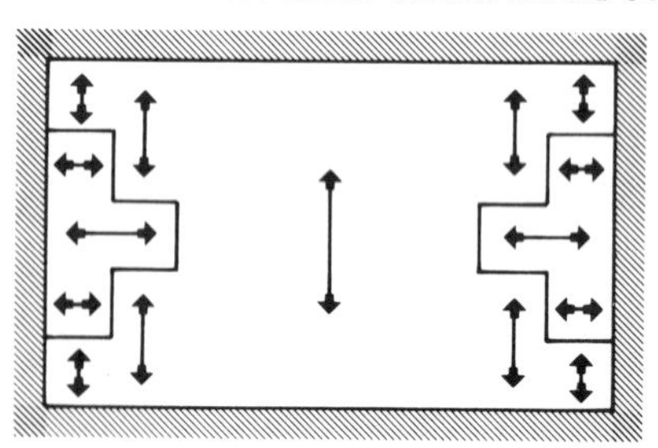

Figure 8.31

8.3.3 Correction Factor, K

Figure 8.32a shows a plan view of a typical strip, with uneven loading.

$$\text{total load at each end of the strip (end reaction)} = qb\,\frac{(l_1 + l_2)}{2}$$

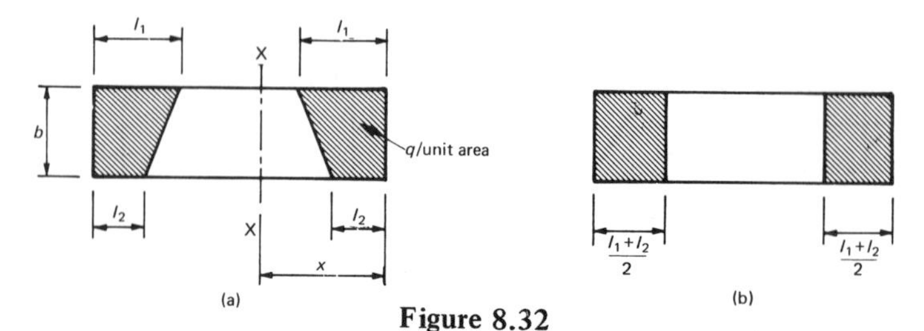

Figure 8.32

To find the position of the centroid of the load from the end of the beam

$$qb\,\frac{(l_1+l_2)}{2}\times\bar{x} = qb\times l_2\times\frac{l_2}{2} + qb\,\frac{(l_1-l_2)}{2}\times\left(l_2+\frac{l_1-l_2}{3}\right)$$

$$= qb\left[\frac{l_2^2}{2} + \frac{(l_1-l_2)}{2}\,\frac{(2l_2+l_1)}{3}\right]$$

$$= \frac{qb}{6}\,(3l_2^2 + 2l_1l_2 + l_1^2 - 2l_2^2 - l_1l_2)$$

$$= \frac{qb}{6}\,(l_1^2 + l_1l_2 + l_2^2)$$

$$\bar{x} = (l_1^2 + l_1l_2 + l_2^2)/3(l_1+l_2)$$

hence the moment at section X–X is

$$M = qb\,\frac{(l_1+l_2)}{2}\,x - qb\,\frac{(l_1+l_2)}{2}\,(x-\bar{x})$$

$$= qb\,\frac{(l_1+l_2)}{2}\,\bar{x}$$

(the moment at any section where $x > \bar{x}$ is constant)

$$M = \frac{qb}{6}\,(l_1^2 + l_1l_2 + l_2^2)$$

Figure 8.32b shows the same strip with the load spread evenly across the width of the strip. At X–X

$$M = \frac{qb(l_1+l_2)^2}{8}$$

so that

$$Kqb\,\frac{(l_1+l_2)^2}{8} = \frac{qb(l_1^2 + l_1l_2 + l_2^2)}{6}$$

$$K = \frac{4}{3}\,\frac{(l_1^2 + l_1l_2 + l_2^2)}{(l_1+l_2)^2}$$

$$= \frac{4}{3}\left[\frac{(l_1+l_2)^2 - l_1l_2}{(l_1+l_2)^2}\right]$$

$$= \frac{4}{3}\left(1 - \frac{l_1l_2}{l_1^2 + 2l_1l_2 + l_2^2}\right)$$

$$K = 1.333 - \frac{1.333}{\dfrac{l_1}{l_2} + 2 + \dfrac{l_2}{l_1}}$$

Returning to the example in the previous section, in strips 2 and 3 $l_1 = L_1/2$ and $l_2 = 0$

$$K = 1.333 - \frac{1.333}{\frac{L_1/2}{0} + 2 + \frac{0}{L_1/2}} = 1.333$$

and

$$\text{design moment for both strips} = 1.333 \frac{qL_1^2}{32} = \frac{qL_1^2}{24}$$

8.3.4 Fixed Edge Slab

A slab with three fixed edges and one simply supported edge is shown in figure 8.33. The method can be applied to the non-rectangular slab with no problems. A fixed edge is more rigid than a simply supported edge and would therefore attract more of the load in the corner. This has been allowed for in this slab, the line of discontinuity runs nearer to the simply supported edge than the fixed edge.

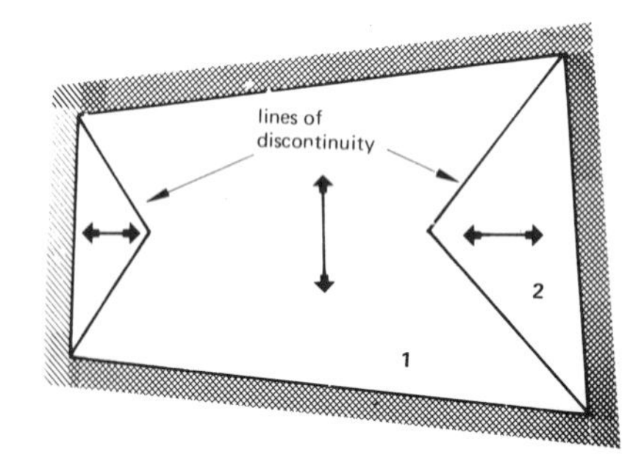

Figure 8.33

The free and reactant BM method is most convenient for finding the design moments.

Strip 1

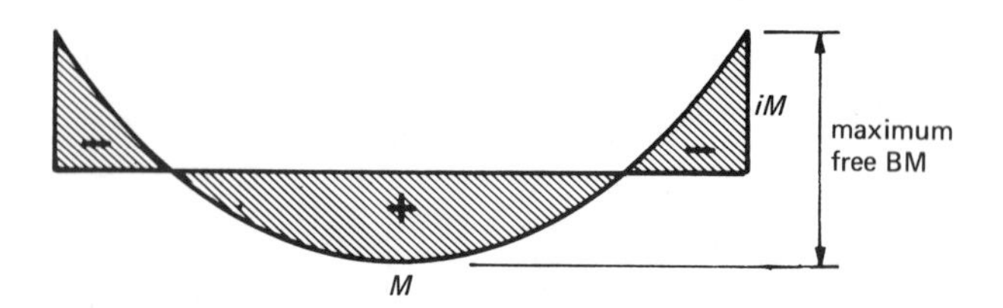

$(1 + i) M =$ maximum free BM

Strip 2

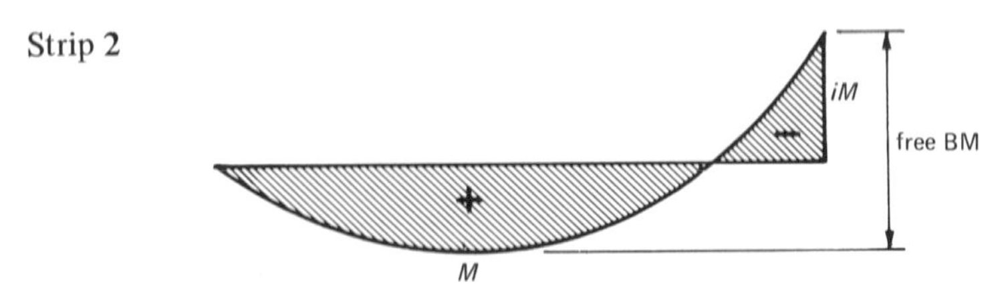

The strips are standard cases which were considered in chapter 3. The relative values of the edge and mid-slab moments (effectively the factor i) must be decided on the same practical basis as in the yield line method (section 8.2.9).

8.3.5 Slab with a Hole

The presence of a hole effectively destroys the continuity of the slab, and it is impossible for strips to span from one side of the slab to the other. This can be overcome by using 'strong strips' around the hole, as shown in figure 8.34. These strips provide the support for some of the other strips, and also reinforce the surround of the hole. This can be seen by looking at the loading on each strip.

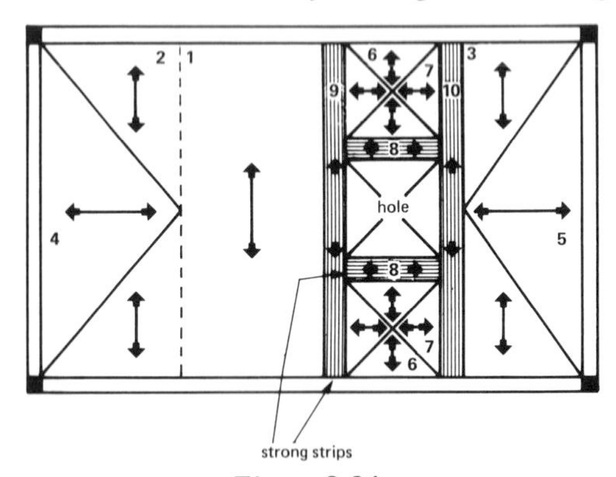

Figure 8.34

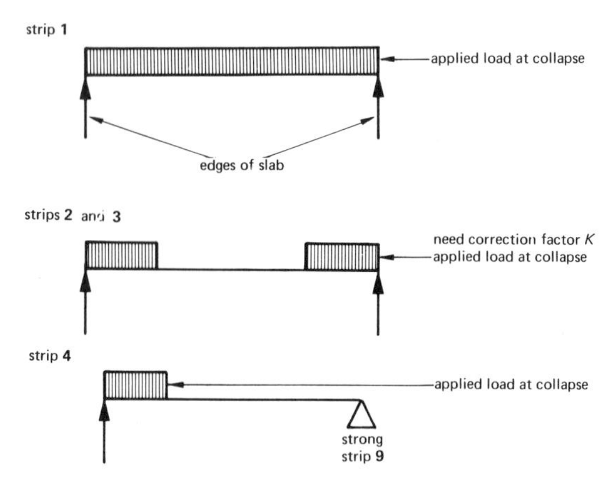

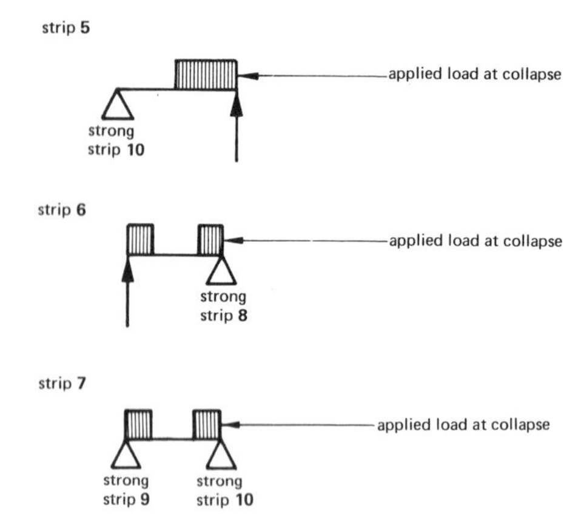

So far, the calculations have proceeded along the same lines as in the previous examples. The loading on the strong strips is more complicated. It is necessary to choose a value for the width of the strong strips (500 mm is a reasonable first guess).

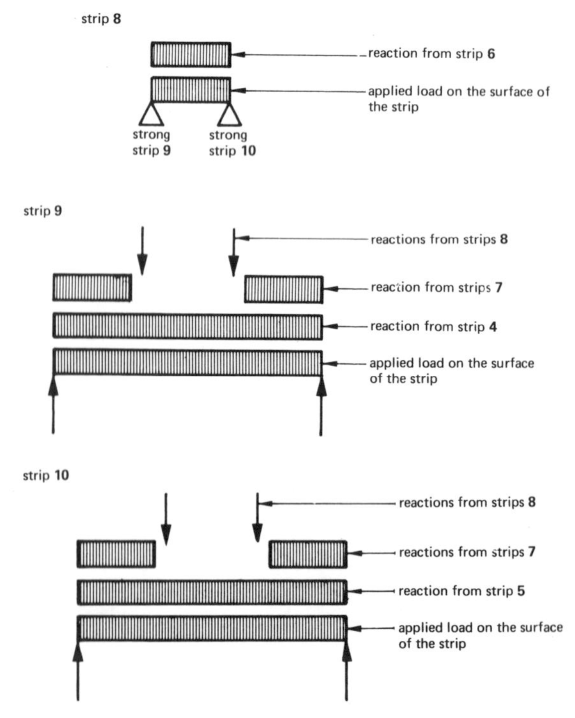

It is worth looking at a few points about the example.

(1) All the applied load on the slab eventually is supported by the edges of the slab, although some strips are supported by other strips.

(2) It is assumed that the reactions at the ends of a strip are constant across the width of the strip, whatever the plan shape of the loading on the strip.

(3) Care is required when considering the strong strips. In ordinary strips it is convenient to use a metre width of strip, but strong strips are usually less than a metre wide!

8.3.6 Loading on Edge Beams

It is usually difficult to assess the distribution of loading on the edge beams of a slab. Hillerborg's strip method provides a neat solution to this problem. The reactions at the ends of strips which are supported by the edge beams are the loads on the beams. Figure 8.35 shows roughly the distributions which would have arisen in the example in the previous section. The distributions assume that the reaction is constant across the width of the strips. That assumption is obviously untrue for strip **4** and it would seem prudent to assume the more realistic distribution indicated by the broken lines.

It is a good idea to check that the total load on all the edge beams is equal to the load applied to the slab. It is the only independent check which can be made of the individual strip calculations.

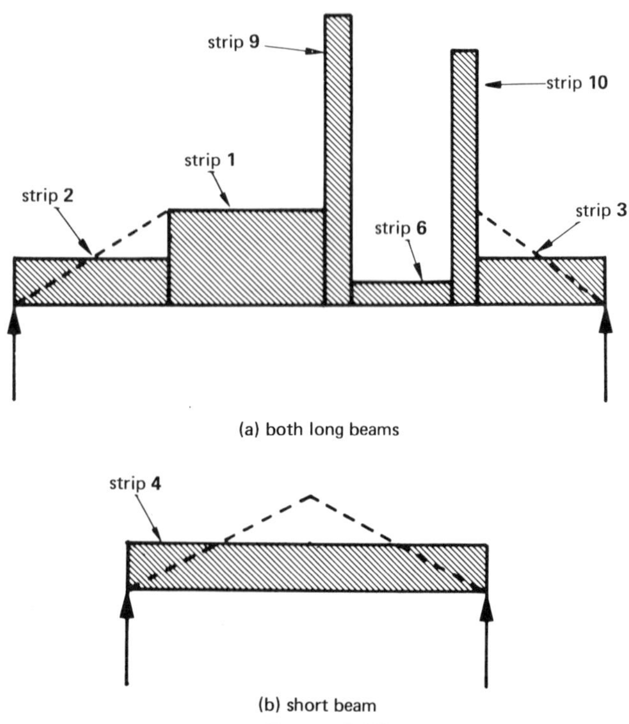

Figure 8.35

8.4 SUMMARY

This final chapter has examined the yield line and Hillerborg strip methods for reinforced concrete slabs. Both methods are based on plastic theory, but because of the inherent complexity of slab behaviour they do not give exact solutions. Yield line theory gives a genuine upper bound within the limitations of its initial assumptions, the strip method manufactures a lower bound. Membrane action and strain hardening of the reinforcement ensure that both methods are safe.

In the yield line method a collapse mechanism is postulated and analysed by the virtual work method (or the equilibrium method). It is necessary to determine the critical arrangement of yield lines in a given mechanism and also to consider alternative mechanisms. Yield line theory can cater for slabs with point loads as well as distributed loading.

In the strip method the loading on the slab is distributed to beam strips spanning between opposite edges of the slab. The method is very flexible when considering slabs with distributed loading but cannot deal easily with point loading. [35] It gives information about the extra reinforcement around holes in the slab and about the loading on the edge beams.

8.5 PROBLEMS

8.1 Determine the collapse load (UDL) of a rectangular RC slab (5 m x 3 m), simply supported on all sides, when the reinforcement is isotropic with $M = 10$ kN m/m. Design the slab to carry twice this load, assuming the edges are now fixed.

8.2 Determine the collapse load and mechanism of the slabs shown in figures 8.36 and 8.37

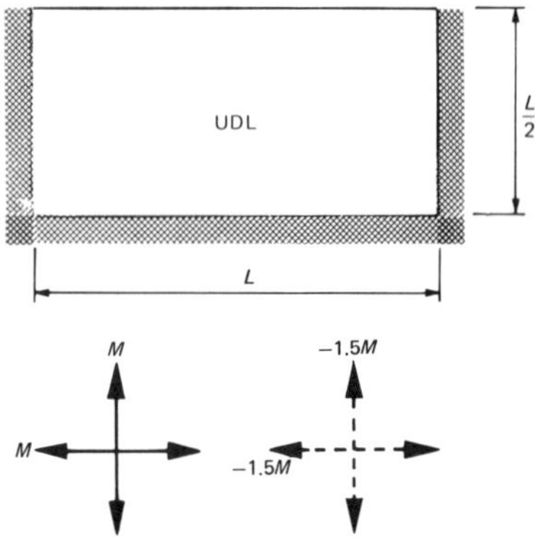

Figure 8.36

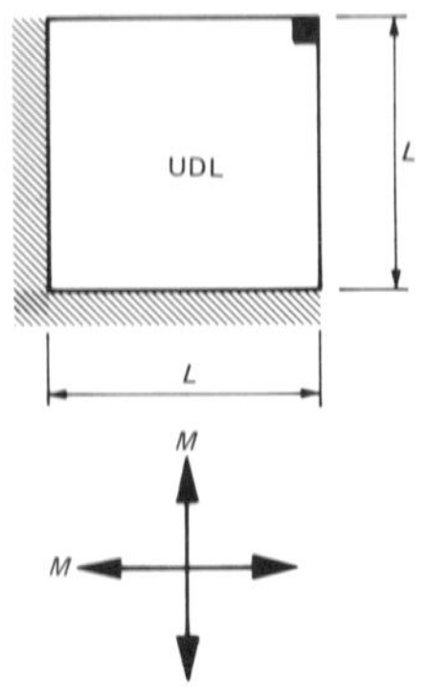

Figure 8.37

8.3 A rectangular slab (7 m x 4 m) has three simply supported and one clamped edge and carries a uniform load q. Figure 8.38 gives details of the reinforcement and yield line pattern. Use the work equation to show that

$$q_c = \frac{3M}{(42 - 4x)}\left(\frac{17.5}{y} + \frac{7}{4 - y} + \frac{4}{x}\right)$$

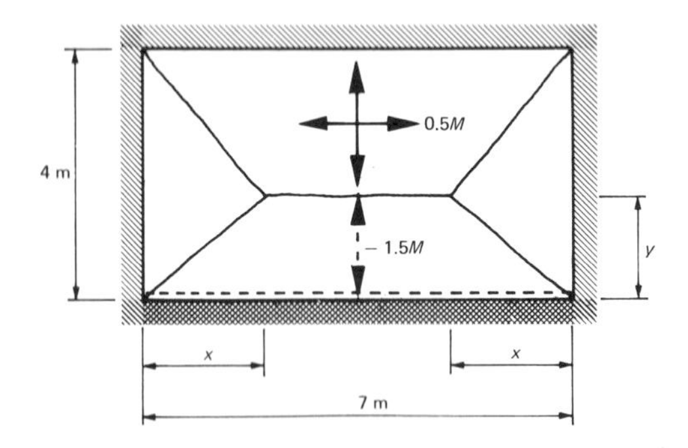

Figure 8.38

Find the moment of resistance M when $q = 9$ kN/m^2. What are the critical values of x and y?

8.4 Determine the relationship beween q, M and L for the yield line system shown in figure 8.39

8.5 Figure 8.40 shows a proposed yield line mechanism for a clamped square slab carrying a UDL q.

(a) Show that the work equation gives

$$q\,[(L - 2R)(L^2 + 2LR + 4R^2) + 2\pi R^3] = 48M\left(L - 2R + \frac{\pi R}{2}\right)$$

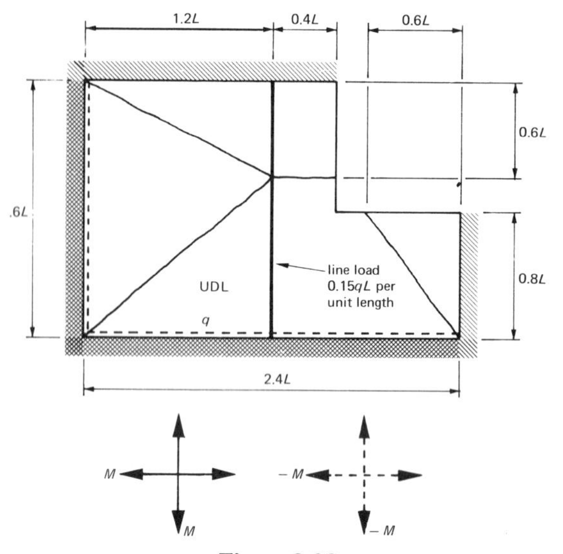

Figure 8.39

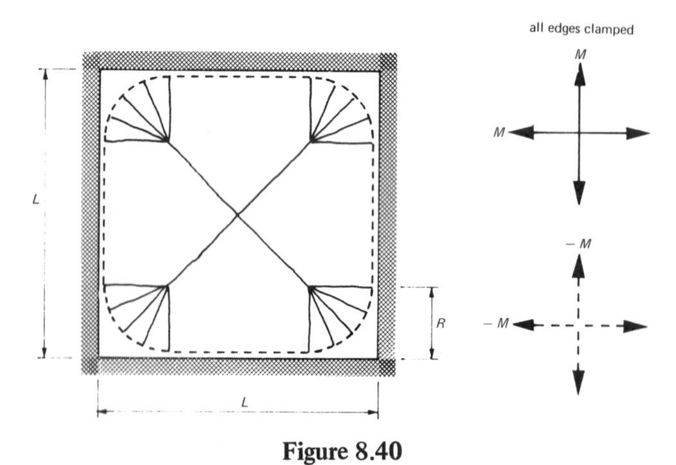

Figure 8.40

(b) Show by a graphical method, or otherwise, that the critical value of R is 0.3L.

(c) Find the collapse load q_c in terms of M and L
(Note: the centroid of a segment of a circle is (2/3) x radius from the centre of the circle).

8.6 A rectangular slab ($3L \times 2L$) is supported by edge beams. The slab has isotropic reinforcement in the bottom with a moment of resistance M per unit length. The edge beams all have a moment of resistance of $4ML$. Determine the collapse mechanism and load when the slab carries a UDL q.

The moment of resistance of one of the long beams is reduced to $2ML$. What effect does this have on the collapse load?

8.7 Use Hillerborg's strip method to design a simply supported slab (10 m x 6 m) to carry a collapse load of 12 kN/m^2.

8.8 Redesign the slab in the previous example to carry a load of 30 kN/m^2 with clamped edges.

8.9 Redesign the same slab with one long edge clamped, the other long edge unsupported and both short edges simply supported, to carry a load of 12 kN/m^2.

8.10 Sketch a suitable system of strips for designing the slab in figure 8.39 by Hillerborg's method.

8.11 Determine the design moments for the slab in figure 8.41 by the strip method. What are the design moments in the edge beams? Design load (including load factor) = 8 kN/m^2. Strong strips around the hole should be 0.5 m wide.

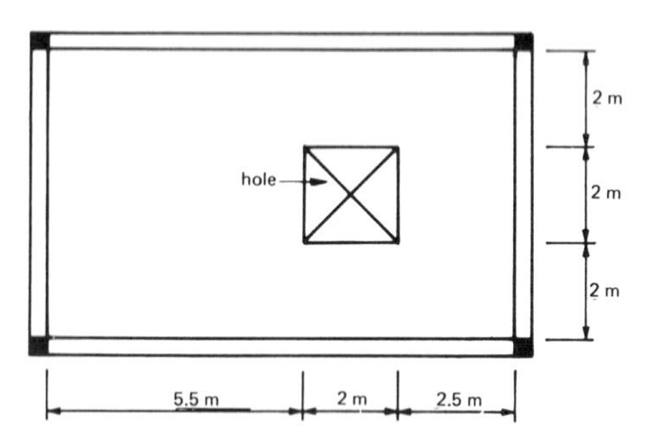

Figure 8.41

APPENDIX A YIELD CRITERIA

In a tensile test a ductile material will yield when the applied stress reaches a critical value. In the case of steel that value is the yield stress. (Other materials may not have a distinct yield point, in which case the sitution is less clear cut.) In a practical situation there may be several stresses in the material, for example direct stress due to bending and shear stress, and the onset of plastic behaviour is more difficult to predict. Various 'yield criteria' have been devised to do this.

Modern materials science has shown that plastic flow in a ductile crystalline material such as a metal is a shearing action within the lattice of atoms which make up the crystal, due to dislocations within the lattice. [36] As might be expected, the two yield criteria, named after Tresca and Von Mises, which assume that shear controls yield, give the most accurate predictions of the onset of yield in ductile metals. These criteria will be examined in some detail.

THE TRESCA CRITERION

This states that yield will occur when the maximum shear stress due to a combination of individual stresses becomes equal to the maximum shear stress which occurs in a simple tensile test on the same material.

Any two-dimensional system of stresses at a given point in the material can be plotted on a Mohr's circle of stress, as in figure A.1a and b. [37]
The maximum and minimum direct stresses at the point, σ_1 and σ_2 (called the principal stresses), and the maximum shear stress τ_{max} can be found from the geometry of the circle. The stresses in figure A.1 are acting in a *z-x* plane. If they are part of a three-dimensional system of stresses it is possible to define all the stresses by means of three Mohr's circles. If the three Mohr's circles are plotted together they turn out not to be independent of each other, as shown in figure A.2. This is because the stresses are all related by the requirements of equilibrium. The three circles define explicitly the stresses in any direction. The maximum shear stress at the point is the radius of the largest Mohr's circle.

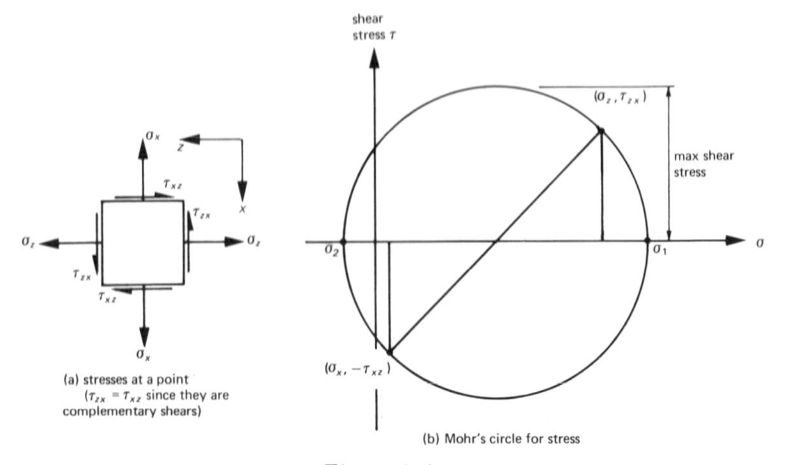

(a) stresses at a point ($\tau_{zx} = \tau_{xz}$ since they are complementary shears)

(b) Mohr's circle for stress

Figure A.1

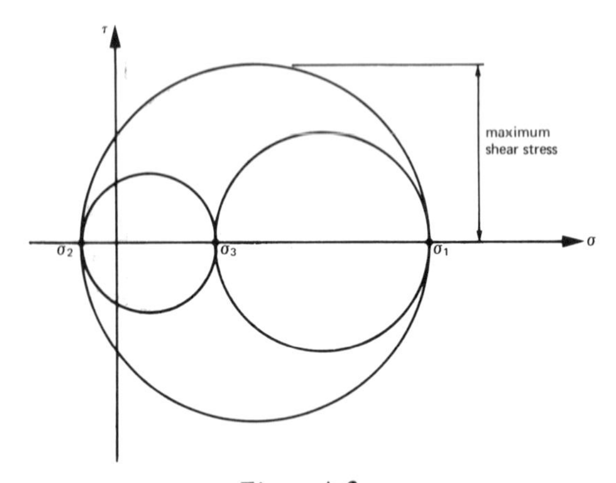

Figure A.2

In a two-dimensional stress system, such as a beam in bending, the third principal stress must be zero. Using the Mohr's circles in figure A.2 it is possible to find the resulting maximum shear stress.

When $\sigma_1 > \sigma_2$ but both have the same sign

$$\tau_{\max} = \frac{\sigma_1 - 0}{2} = \frac{\sigma_1}{2} \tag{A.1}$$

When $\sigma_1 > \sigma_2$ but they are of opposite sign

$$\tau_{\max} = \frac{\sigma_1 - \sigma_2}{2} \tag{A.2}$$

In the tensile test there is a one-dimensional system of stress, so that $\sigma_2 = \sigma_3 = 0$.

At yield the other principal stress is the yield stress σ_y. Thus the yield shear stress τ_y is

$$\tau_y = \frac{\sigma_y - 0}{2} = \frac{\sigma_y}{2} \tag{A.3}$$

Equating equations A.1 and A.2 to equation A.3 gives the conditions for yield to occur

$$\begin{aligned} &\sigma_1 = \sigma_y \text{ or } 2\tau_y \qquad \sigma_1 > \sigma_2 \text{ both same sign} \\ &\sigma_1 - \sigma_2 = \sigma_y \text{ or } 2\tau_y \qquad \sigma_1 > \sigma_2 \text{ opposite signs} \end{aligned} \tag{A.4}$$

VON MISES CRITERION

In an elastic material with a two-dimensional system of stress the strain energy per unit volume [37] in terms of the principal stresses is

$$U = \frac{1}{2}(\sigma_1\epsilon_1 + \sigma_2\epsilon_2) \tag{A.5}$$

Using the generalised form of Hooke's law

$$\begin{aligned} \epsilon_1 &= \frac{1}{E}(\sigma_1 - v\sigma_2) \\ \epsilon_2 &= \frac{1}{E}(-v\sigma_1 + \sigma_2) \end{aligned} \tag{A.6}$$

Substituting equation A.6 into equation A.5 gives

$$\begin{aligned} U &= \frac{1}{2E}[\sigma_1(-v\sigma_2) + \sigma_2(-v\sigma_1 + \sigma_2)] \\ &= \frac{1}{2E}(\sigma_1^2 + \sigma_2^2 - 2v\sigma_1\sigma_2) \\ &= \frac{(1-2v)}{6E}(\sigma_1 + \sigma_2)^2 + \frac{(1+v)}{6E}(2\sigma_1^2 + 2\sigma_2^2 - 2\sigma_1\sigma_2) \end{aligned}$$

Remembering that the shear modulus $G = E/2(1 + v)$

$$U = \frac{(1-2v)}{6E}(\sigma_1 + \sigma_2)^2 + \frac{1}{6G}(\sigma_1^2 + \sigma_2^2 - \sigma_1\sigma_2) \tag{A.7}$$

Volumetric strain energy per unit volume — Shear strain energy per unit volume, U_{SSE}

It can be shown [37] that the first part of equation A.7 is the strain energy due to the change in volume caused by the stresses, and the second part, called the *shear strain energy* per unit volume is the strain energy due to the change in shape causes by the shearing action of the stresses.

The Von Mises criterion states that the material will yield when the shear strain energy per unit volume caused by the applied stresses is equal to the shear strain energy per unit volume at yield in a tensile test.
In the tensile test

$$\sigma_1 = \sigma_y$$

$$\sigma_2 = 0$$

$$U_{sse} = \frac{\sigma_y{}^2}{6G}$$

so that yield will occur when

$$\sigma_1^2 + \sigma_2^2 - \sigma_1\sigma_2 = \sigma_y{}^2 \tag{A.8}$$

In a situation of pure shear, yield occurs when the shear stress is equal to the yield shear stress τ_y. The Mohr's circle for pure shear is shown in figure A.3. From the geometry of the circle

$$\sigma_1 = \tau_y$$

$$\sigma_2 = -\tau_y$$

Using the criterion, yield will occur when

$$(\tau_y)^2 + (-\tau_y)^2 - (\tau_y)(-\tau_y) = \sigma_y{}^2$$

$$3\tau_y{}^2 = \sigma_y{}^2 \tag{A.9}$$

The predictions of both criteria are summarised in equations A.4 and A.8. It is generally recognised that the Von Mises criterion predicts more accurately

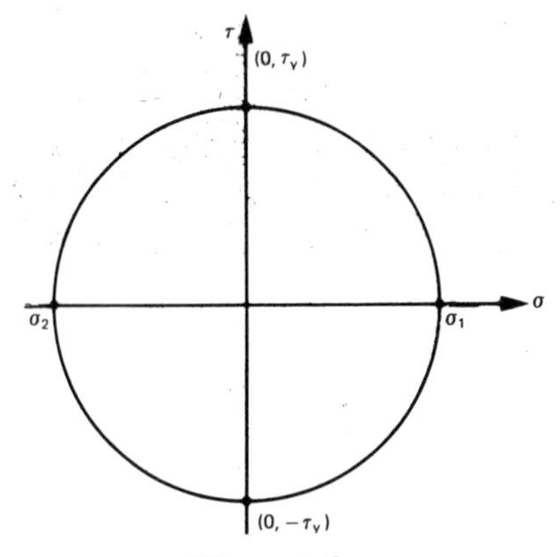

Figure A.3

when yield will occur in a ductile material. In chapter 2 both criteria were used to predict yield in a beam subject to direct stresses σ from bending and shear stresses τ. The direct stresses are parallel to the axis of the beam, so that the stresses are as in figure A.1a with

$\sigma_x = \sigma$

$\sigma_z = 0$ no direct stresses normal to the beam axis

$\tau_{xz} = \tau$

The Mohr's circle for these stresses is shown in figure A.4. From the geometry of the circle

$$\sigma_1 = \frac{\sigma}{2} + \sqrt{\left(\frac{\sigma^2}{4} + \tau^2\right)}$$
$$\sigma_2 = \frac{\sigma}{2} - \sqrt{\left(\frac{\sigma^2}{4} + \tau^2\right)} \qquad \text{(A.10)}$$

Equation A.10 can be substituted into equations A.4 and A.8 to compare when the two criteria predict yield will occur. As table A.1 shows, both criteria predict that yield will occur in the beam when

$$\left(\frac{\sigma}{\sigma_y}\right)^2 + \left(\frac{\tau}{\tau_y}\right)^2 = 1 \qquad \text{(A.11)}$$

The only difference between them is the value of τ_y, the value of the yield shear stress.

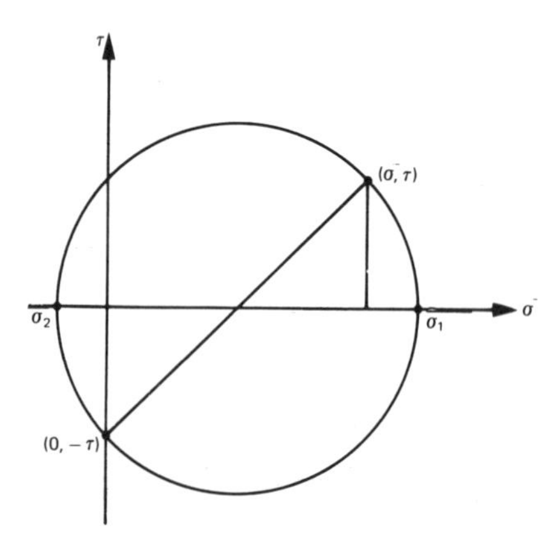

Figure A.4

Table A.1

Tresca	Von Mises
From equation A.10, $\sigma_1 > \sigma_2$ but they are of opposite sign, hence yield occurs when	Substituting equation A.10 into equation A.8
$2\sqrt{\left(\frac{\sigma^2}{4}+\tau^2\right)}=\sigma_y$	$\left[\frac{\sigma}{2}+\sqrt{\left(\frac{\sigma^2}{4}+\tau^2\right)}\right]^2+\left[\frac{\sigma}{2}-\sqrt{\left(\frac{\sigma^2}{4}+\tau^2\right)}\right]^2$ $-\left[\frac{\sigma}{2}+\sqrt{\left(\frac{\sigma^2}{4}+\tau^2\right)}\right]\left[\frac{\sigma}{2}-\sqrt{\left(\frac{\sigma^2}{4}+\tau^2\right)}\right]$ $={\sigma_y}^2$
$\sigma^2+4\tau^2={\sigma_y}^2$	$\frac{\sigma^2}{2}+2\left(\frac{\sigma^2}{4}+\tau^2\right)-\frac{\sigma^2}{4}$ $+\left(\frac{\sigma^2}{4}+\tau^2\right)={\sigma_y}^2$
or $\left(\frac{\sigma}{\sigma_y}\right)^2+\left(\frac{2\tau}{\sigma_y}\right)^2=1$	$\sigma^2+3\tau^2={\sigma_y}^2$
from equation A.4 $\sigma_y=2\tau_y$ so that yield occurs when	or $\left(\frac{\sigma}{\sigma_y}\right)^2+3\left(\frac{\tau}{\sigma_y}\right)^2=1$
$\left(\frac{\sigma}{\sigma_y}\right)^2+\left(\frac{\tau}{\tau_y}\right)^2=1$	From equation A.9 $3{\tau_y}^2={\sigma_y}^2$
$\tau_y=0.5\sigma_y$	so that yield will occur when $\left(\frac{\sigma}{\sigma_y}\right)^2+\left(\frac{\tau}{\tau_y}\right)^2=1$
	$\tau_y=0.577\sigma_y$

APPENDIX B A REDUNDANCY TEST

Most text books on structural analysis contain various tests for redundancy. In this book it has only been necessary to find the degree of redundancy of two-dimensional frames, and it is possible to give a specific test for this.

As was discussed in the early part of the book there are several ways of thinking about redundancy. Another way is to define the degree of redundancy as the number of cuts (often called releases) which must be made in the structure to make it statically determinate. In two-dimensional frames with rigid joints and supports the columns can be made statically determinate cantilevers by breaking the structure at the centre of every span, as in figure B.1. Each break is effectively three cuts, because it allows relative horizontal and vertical displacements and rotation.

A frictionless hinge (at a support or within the frame) or a roller bearing at a support, as in figure B.2, reduces the degree of redundancy by one. Thus the test for redundancy is

$$r = 3n - k$$

where r = degree of redundancy n = number of spans, k = number of frictionless hinges or roller bearings.

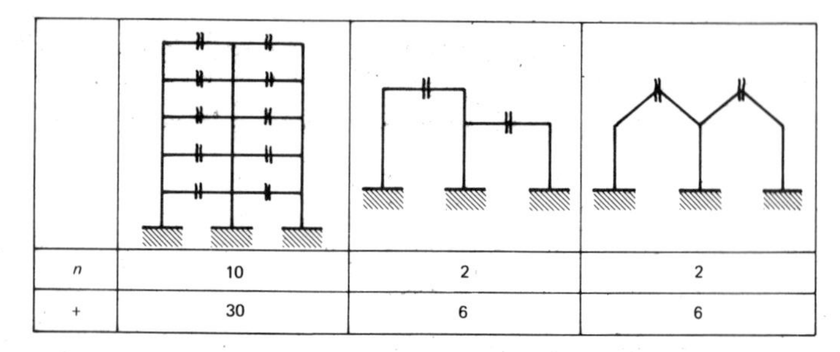

n	10	2	2
+	30	6	6

Figure B.1

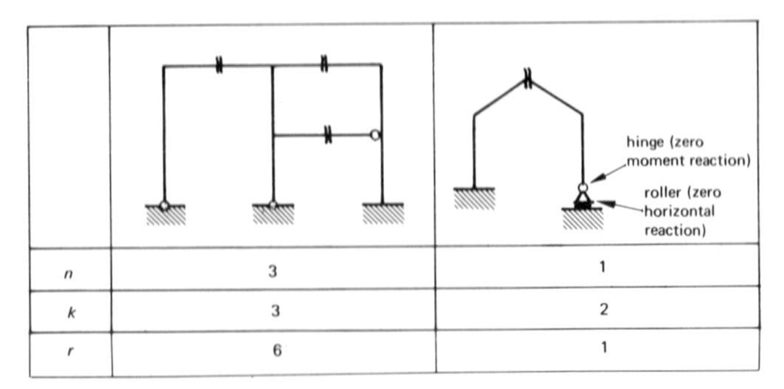

n	3	1
k	3	2
r	6	1

Figure B.2

A warning about finding the degree of redundancy. Consider the joint in figure B.3a. There are bending moments at the end of each member but only two of them are independent, the third can be found from moment equilibrium of the joint. Consequently, the pin-joint shown is equivalent to frictionless hinges at the ends of *two* of the three members. As shown in figure B.3b this can be extended to joints of more than three members.

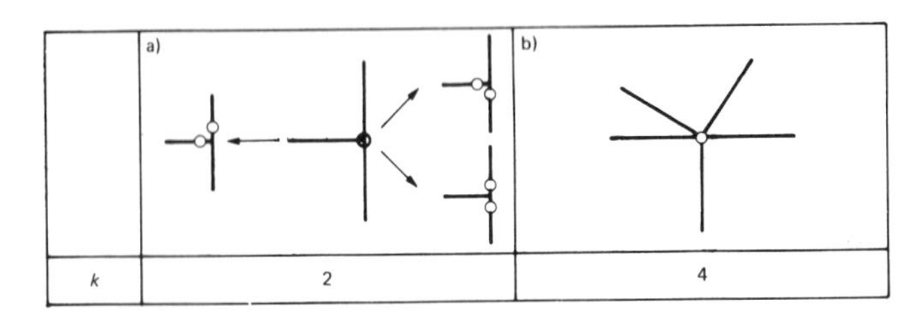

	a)	b)
k	2	4

Figure B.3

APPENDIX C BENDING MOMENT DIAGRAMS

C.1 INTRODUCTION

It is often necessary to determine the BMD corresponding to a collapse mechanism. Usually the BMD is required to check that the mechanism satisfies the yield condition and is thus the true collapse mechanism of the structure. The object of this appendix is to help the reader to find the BMD, even when the structure is quite complicated. Sections C.2 to C.5 give some useful ideas on this, and section C.6 shows how they are applied in an example.

C.2 INFORMATION FROM THE MECHANISM ITSELF

At the point of collapse the structure is statically determinate. This is due to the formation of plastic hinges. At every plastic hinge the magnitude of the BM is known – it is equal to the plastic moment of the weakest member at that point. The *sense* of the moment is also known. Consider the mechanism in figure C.1. At the hinge points A, C, D and E the moment is equal to M_p. Consider now points A and E in more detail. The members are not just the lines shown in figure C.1, they have finite depth and width. At a plastic hinge the material is yielding in tension and compression. The areas in tension and compression at A and E must be as in figure C.2 to allow the plastic rotations required in the mechanism.

Similar arguments can be used at C and D, as shown in figures C.3a and b. The magnitude and sense of the BM at every hinge can now be plotted. (The author's

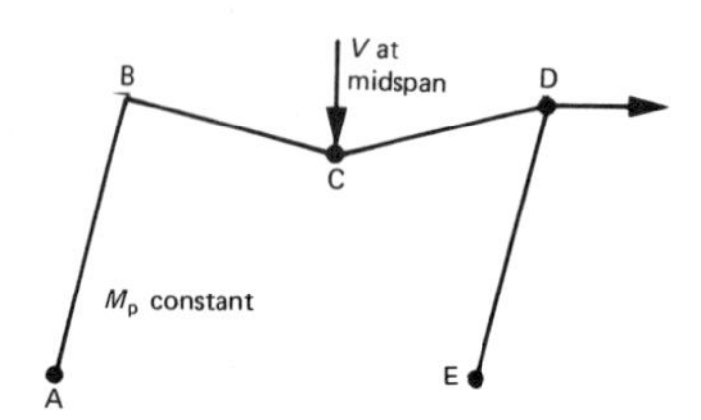

Figure C.1

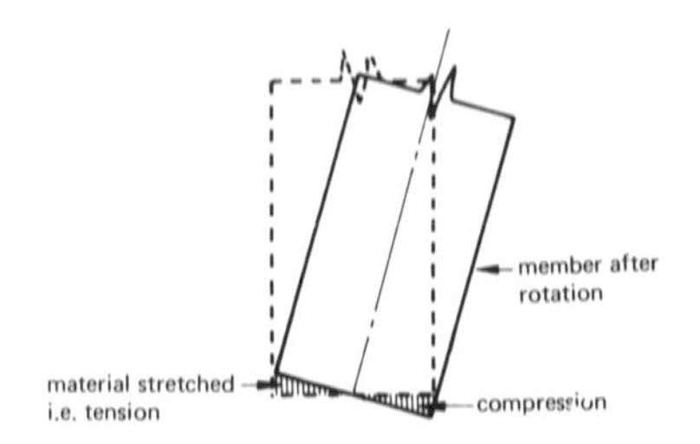

Figure C.2

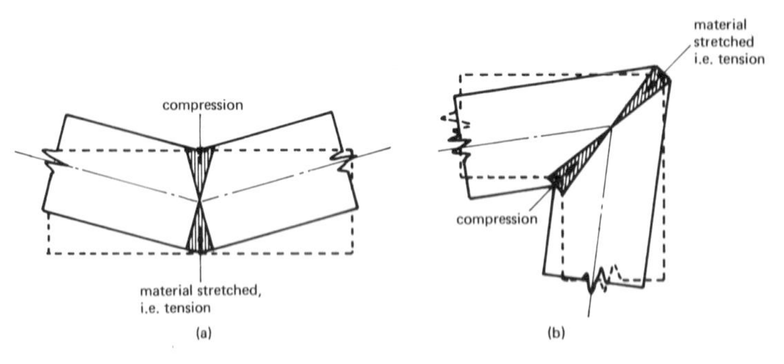

Figure C.3

preference is to plot the moment on the side of the member which is in tension. This convention has been followed in all the BMDs in the book.)

C.3 FREE AND REACTANT BENDING MOMENTS

It is convenient to divide any BMD into two parts. These are the *free* and *reactant* BMDs. The free BMs in a member are the BMs which would be caused by loads applied to the member if the ends of the member were free to rotate; in other words, the BMs in a simply supported beam of the same length as the member.

The reactant moments are the BMs at the ends of the member due to restraint of the ends against rotation from the rest of the structure. (The object of many elastic analyses is to find the reactant moments.) Connecting the end moments by a straight line gives the reactant BMD.

There are simple relationships between the geometries of the free and reactant BMDs and the geometry of the actual BMD, which are very useful in plastic analysis. (The free and reactant BMD method described in chapter 3 is based entirely on this.) Look at the beam BD in figure C.1. The BMD for the beam is going to be as in figure C.4a, but the reactant moment at B is not known. Divide the BMD into the free and reactant parts as in figure C.4b. Under the point load, which is at midspan in this case, the geometry of the diagrams requires that

$$\begin{array}{l}\text{Maximum}\\ \text{free BM}\end{array} - \frac{M_p + M_B}{2} = M_p$$

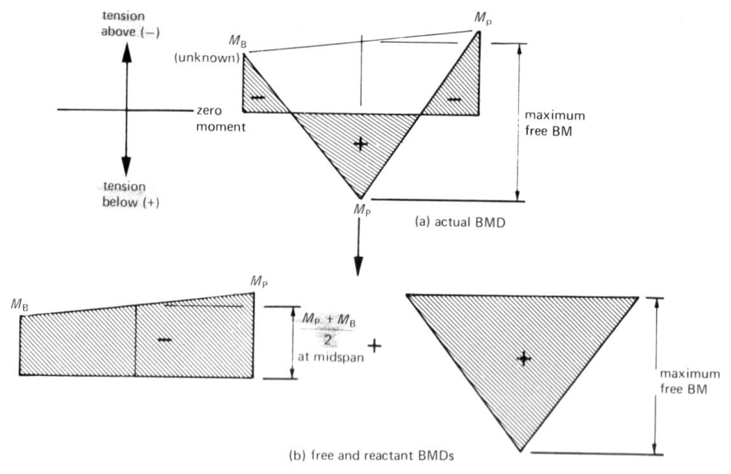

Figure C.4

For a simply supported beam, span L, with a central point load V, the maximum free BM is $VL/4$ so that

$$\frac{VL}{4} - \frac{M_p + M_B}{2} = M_p$$

$$M_B = \frac{VL}{2} - 3M_p$$

If M_B comes out as a positive number then the guess of tension on the top of the beam at B, as in figure C.5 is correct. If it comes out negative, the guess is incorrect, the tension is on the underside of the beam.

There is one other point which causes consternation in the combination of the free and reactant BMs. The calculations only make use of the *magnitudes* of

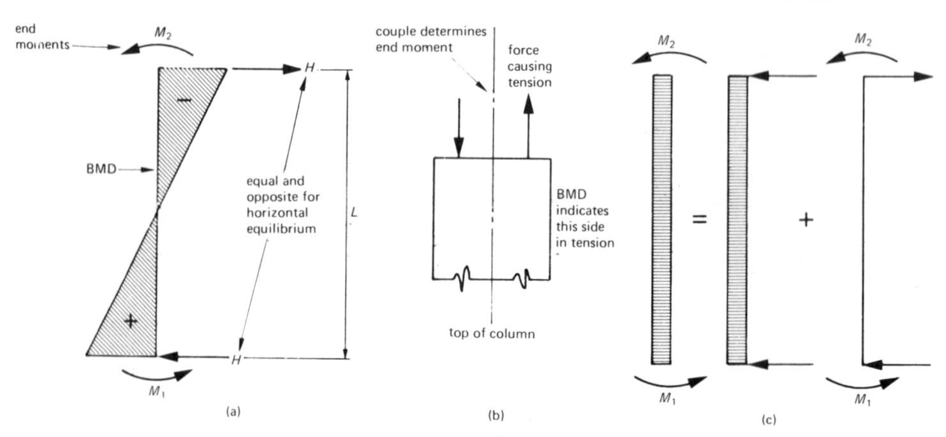

Figure C.5

the moments. Thus, in the calculations based on figure C.4, only vertical distances representing these magnitudes were considered. In order to preserve these vertical distances when the free and reactant BMDs are combined, the sloping sides of the free BMD have to be distorted. They remain straight lines but their length is changed. The process is entirely consistent.

C.4 END FORCES ON MEMBERS

Consider the column BMD shown in figure C.5a. The actual BMD is the same as the reactant BMD indicating that there is no applied load along the length of the column. The end moments from the rest of the structure which cause the reactant BMs are also shown. (Figure C.5b shows a convenient way of finding the direction of the end moment.)

The column must be in equilibrium. The only way that moment equilibrium about the top (or base) of the column can be achieved is to have equal and opposite horizontal forces acting as shown. Taking moments about the top of the column

$$HL = M_1 + M_2$$

These horizontal forces can be thought of as applied forces from the rest of the structure. If the bottom of the column is connected to a support, the horizontal force is the horizontal reaction at the base, just as the end moment is the moment reaction at the base.

When there are loads along the length of the column the horizontal forces are the sum of the forces due to the end moments and the end reactions assuming the column is a simply supported beam, as shown in figure C.5c.

It is often useful to consider horizontal equilibrium of parts of a structure in order to determine parts of the BMD. This will be illustrated in the example in section C.6

C.5 JOINT EQUILIBRIUM

For any structure to be in equilibrium every joint must be in moment equilibrium. This requirement can be very useful when finding BMDs.

A typical joint with parts of the BMDs in each member meeting at the joint is shown in figure C.6a. If the members are cut just beside the joint, the BMs at the ends of the members can be drawn as in figure C.6b. The directions of the moments are found in the manner illustrated in figure C.5b. Moment equilibrium requires that the sum of these moments is equal to zero. Thus, in this case

$$M_1 - M_2 - M_3 = 0$$

C.6 EXAMPLE OF FINDING A BENDING MOMENT DIAGRAM

Limit analysis was used in section 4.4.3 to find the collapse load of the two-storey frame in figure C.7a. The mechanism with lowest load factor is shown in

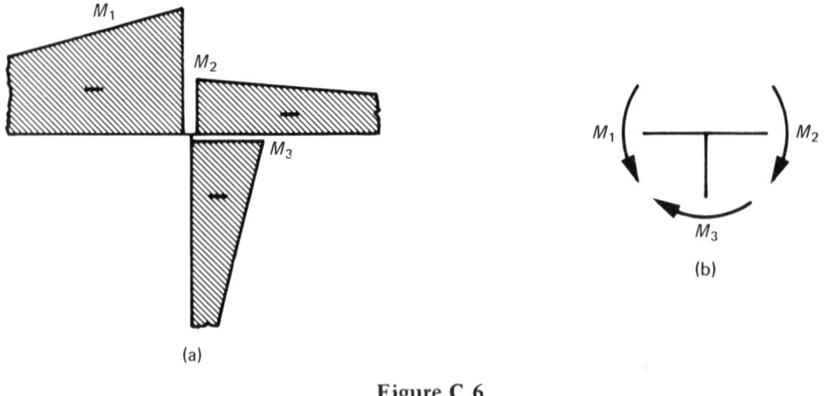

Figure C.6

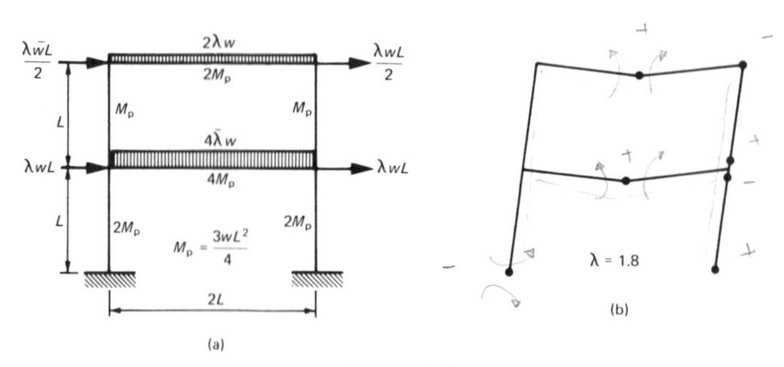

Figure C.7

figure C.7b. It is necessary to check the mechanism by finding the corresponding BMD. The various steps are outlined below and summarised in figure C.8.

The first stage is to mark in all the known moments at the plastic hinges. The BMD for the two right-hand columns can now be completed by joining the end moments by a straight line. The reactant moment at the right-hand end of the lower beam can be found from joint equilibrium. For equilibrium

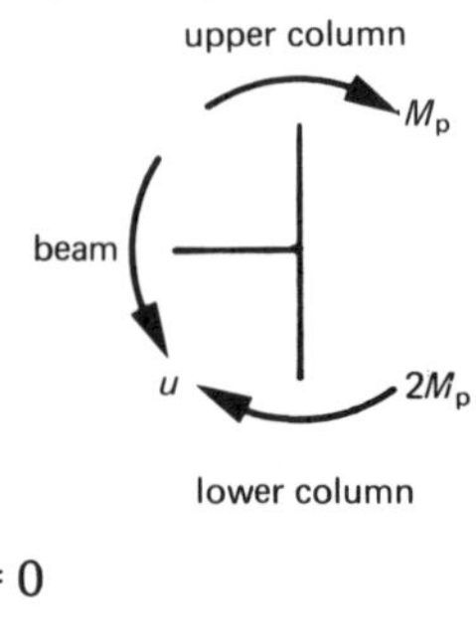

$$u - M_p - 2M_p = 0$$

$$u = 3M_p$$

with tension on the top of the beam.

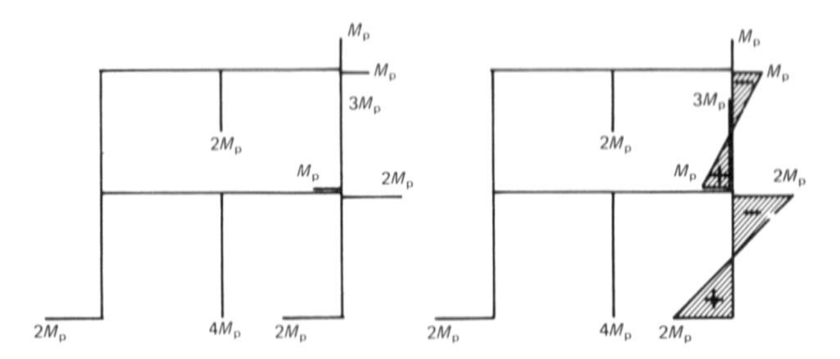

(1) Mark in all known moments at the plastic hinges, and complete the BMD for the right-hand columns.

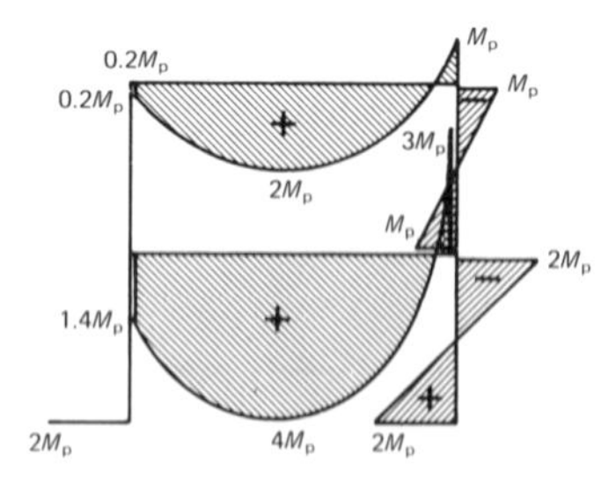

(2) complete the beam BMDs by considering geometry of free and reactant BMDs.

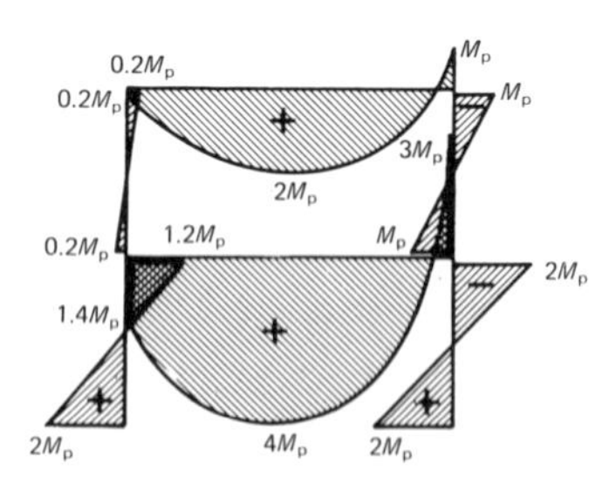

(3) complete the BMD by considering horizontal equilibrium of the upper and lower storeys. Check arithmetic by considering joint equilibrium.

Figure C.8

The unknown moments at the left-hand ends of both beams are found next by considering the geometry of the free and reactant BMDs.

(a) *Upper beam* Free BM at midspan (simply supported beam with UDL)

$$= \frac{2\lambda w\,(2L)^2}{8} = \lambda w L^2 = 2.4M_p$$

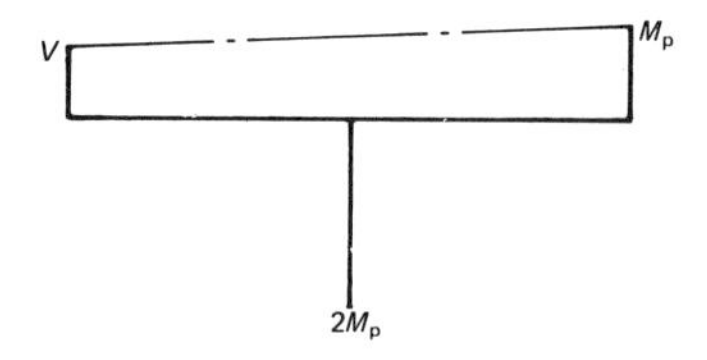

since $\lambda = 1.8$ and $M_p = 3wL^2/4$. Hence from the geometry of the BMDs

$$\frac{M_p + V}{2} + 2M_p = 2.4M_p$$

$$V = -0.2M_p$$

This moment of $0.2M_p$ with tension on the inside can now be plotted and the BMD for the beam completed. Remember that the diagram will be a parabola because of the UDL.

(b) *Lower beam*

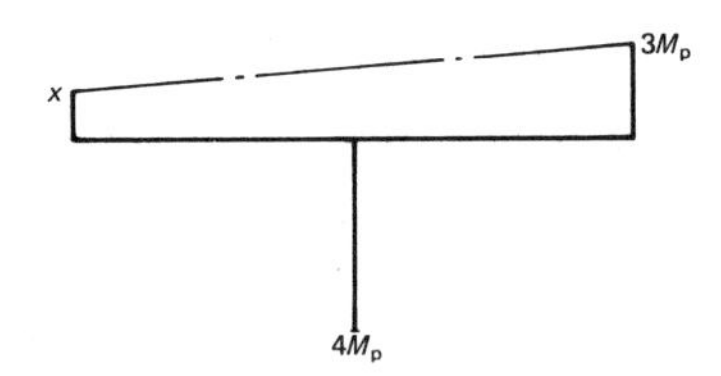

$$\text{Free BM at midspan} = \frac{4\lambda w(2L)^2}{8} = 2\lambda wL^2$$

$$= 4.8M_p$$

and from the geometry of the BMDs

$$\frac{3M_p + x}{2} + 4M_p = 4.8M_p$$

$$x = -1.4M_p$$

This can be used to complete the BMD for the lower beam.

The next step is to consider horizontal equilibrium of both the upper and the lower storey.

(c) *Horizontal equilibrium of the upper storey* This can be achieved by cutting through the bottom of both the upper columns. The diagram shows the

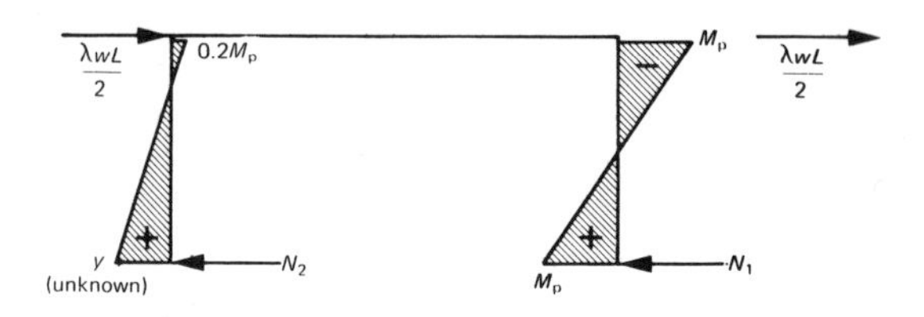

horizontal loads on the upper storey, the shear forces at the base of the columns and the BMs in the columns. For horizontal equilibrium

$$N_1 + N_2 = \lambda wL = 1.8 \times \frac{4M_p}{3L} = \frac{2.4M_p}{L}$$

Using the approach in section C.4

$$N_1 = \frac{2M_p}{L} \qquad N_2 = \frac{0.2M_p + y}{L}$$

so that

$$\frac{0.2M_p + y}{L} + \frac{2M_p}{L} = \frac{2.4M_p}{L}$$

$$y = 0.2M_p$$

This completes the BMD for the upper left-hand column.

(d) *Horizontal equilibrium of the lower storey*

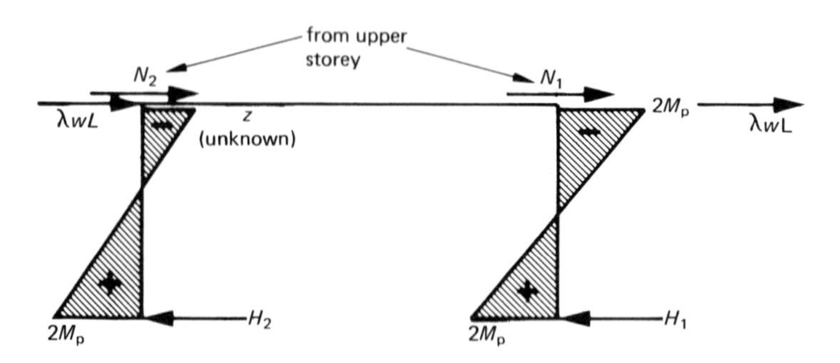

For horizontal equilibrium

$$H_1 + H_2 = N_1 + N_2 + 2\lambda wL$$

$$= \frac{7.2M_p}{L}$$

$$H_1 = \frac{4M_p}{L} \qquad H_2 = \frac{2M_p + z}{L}$$

so that

$$\frac{4M_p}{L} + \frac{2M_p + z}{L} = \frac{7.2M_p}{L}$$

$$z = 1.2M_p$$

This completes the BMD for the lower left-hand column and for the whole structure. However it is worth checking the arithmetic by examining joint equilibrium at the lower left-hand joint.

$$1.2M_p + 0.2M_p - 1.4M_p = 0$$

$$0 = 0$$

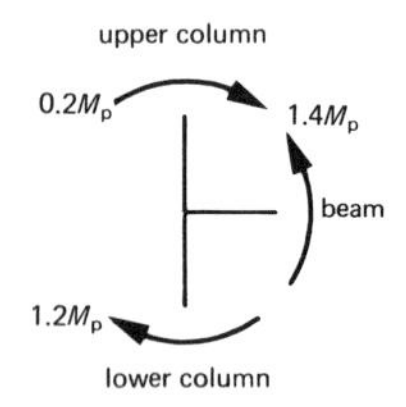

REFERENCES

1. A. Ghali and A. M. Neville, *Structural Analysis*, 2nd ed (Chapman & Hall, London, 1978)
2. *Structural Steelwork Handbook*, new edition (BCSA/Constrado, 1978)
3. M. R. Horne, *Plastic Theory of Structures*, (Nelson, London, 1971)
4. B. G. Neal, *The Plastic Methods of Structural Analysis*, 3rd (SI) ed (Chapman & Hall, London, 1977)
5. CP 110: Part 1: 1972 The Structural Use of Concrete
6 B/20 Draft: Draft Standard Specification for the Structural Use of Steelwork in Building (British Standards Institution, 1978, Draft for Public Comment)
7. BS 449: Part 2: 1969 The Use of Structural Steel in Building
8. *The Collapse Method of Design*, Publication No. 5 (British Constructional Steelwork Association, London, 1952)
9. A. Battersby, *Mathematics in Management* (Penguin, Harmondsworth, 1970)
10. K. I. Majid, *Non-linear structures* (Butterworth, London, 1972)
11. *Genesys Applications Software* (Genesys Limited, Loughborough, Leics., 1978)
12. Engineering Design Programs – Software Archives
13. A. R. Toakley, Optimum design using available sections, *J. Struct. Div., Am. Soc. civ. Engrs,* **94** (1968) 1219–41
14. M. R. Horne and W. Merchant, *The Stability of Frames* (Pergamon, Oxford, 1965)
15. A. C. Walker, *The Buckling of Struts* (Chatto & Windus, London, 1975)
16. R. H. Wood, The stability of tall buildings *Proc. Instn civ. Engrs*, **11** (1958) 69–102
17. R. H. Wood, Effective lengths of columns in multistorey buildings, *BRE Current Paper 85/74* (September 1974)
18. F. K. Kong and R. H. Evans, *Reinforced and Prestressed Concrete* (Nelson, London, 1975)
19. B. P. Hughes, *Limit State Theory for Reinforced Concrete Design*, (Pitman, London, 1976)
20. R. G. Smith, The determination of the compressive stress–strain properties of concrete in flexure, *Mag. Concr. Res.*, **12** (1960) 165–70

21. E. Hognestad, N. R. Hanson and D. McHenry, Concrete stress distribution in ultimate strength design, *J. Am. Concr. Inst.*, **27** (1955) 455–79
22. A. A. Mufti, M. S. Mirza, J. O. McCutcheon and J. Honde, A study of the behaviour of reinforced concrete elements using finite elements, Civil Engineering Report No. 70–5 (Department of Civil Engineering and Applied Mechanics, McGill University, 1970)
23. A. L. L. Baker, *The Ultimate-Load Theory Applied to the Design of Reinforced and Prestressed Concrete Frames* (Concrete Publications Ltd, London, 1956)
24. A. L. L. Baker, Ultimate load design of reinforced and prestressed concrete frames, *Proceedings of a Symposium on the Strength of Concrete Structures* (Cement and Concrete Association, London 1956) 277–304
25. C. E. Massonnet and M. A. Save, *Plastic Analysis and Design*, Vol. 1, Beams and Frames (Blaisdell, London, 1965)
26. K. W. Johansen, *Yield Line Theory* (English translation, Cement and Concrete Association, London, 1962)
27. K. W. Johansen, *Yield Line Formulae for Slabs* (English translation, Cement and Concrete Association, London, 1972)
28. A. J. Ockleston, *Tests on the Old Dental Hospital, Johannesburg* (Concrete Association London, 1956)
29. S. P. Timoshenko and S. Woinowsky-Krieger, *Theory of Plates and Shells*, 2nd edn. (McGraw-Hill Kogakusha Ltd, New York, 1959)
30. R. H. Wood, *Plastic and Elastic Design of Slabs and Plates* (Thames & Hudson, London, 1961)
31. R. H. Wood, *Studies in Composite Construction, Part 2: The interaction of floors and beams in multi-storey buildings* (HMSO, 1961)
32. E. N. Fox, Limit analysis for plates: the exact solution for a clampted square plate of isotropic homogeneous material obeying the square yield criterion and loaded by uniform pressure, *Phil. Trans. R. Soc. A*, **277**, (1974) 121–55
33. L. L. Jones and R. H. Wood, *Yield Line Analysis of Slabs* (Thames & Hudson, London, 1967)
34. A. Hillerborg, *Strip Method of Design* (Viewpoint Publications, Cement and Concrete Association, London, 1975)
35. R. H. Wood and G. S. T. Armer, The theory of the strip method for design of slabs, *Proc. Instn civ. Engrs*, **41**, (1968) 285–311
36. J. E. Gordon, *The New Science of Strong Materials*, 2nd edn, (Penguin, Harmondsworth, 1977)
37. W. A. Nash, *Strength of Materials*, 2nd edn, (Schaum's Outline Series, McGraw-Hill, New York, 1972)
38. *Specification for the Design, Fabrication and Erection of Structural Steel for Buildings* (American Institute of Steel Construction, New York, 1969)

SOLUTIONS TO PROBLEMS

CHAPTER 2

2.1 plates $M_p = Dbt\sigma_y$; combined section $M_p = 660$ kN m

2.3 $M_p = 19.73$ kN m

2.4 (a) $M_p = D^3\sigma_y/6$; (b) $M_p = 1.5d^2t\sigma_y$; (c) $M_p = \sqrt{(2)}d^2t\sigma_y$; (d) $M_p = (7\sqrt{3}/16)\,a^2t\sigma_y$

2.5 $S' = S - \dfrac{A^2}{4d}n^2 \qquad n \leqslant \dfrac{dt_w}{A}$

$$S' = \frac{A^2}{8t_f}(1-n)\left(\frac{4bt_f}{A} - 1 + n\right)$$

y-axis 697.6 kN m, 422.7 kN m; z-axis 122.7 kN m, 109.4 kN m

CHAPTER 3

3.1 $w_c = 11.66M_p/L^2$

3.2 span CD critical; $w_c = 0.833M_p/L^2$

3.3

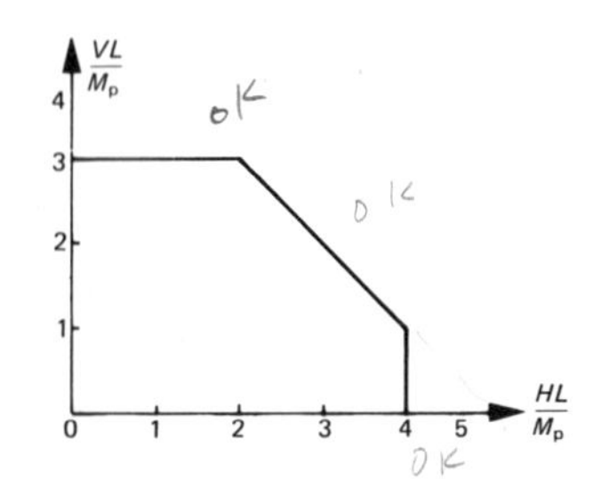

3.4 $\lambda_c = 1.70$

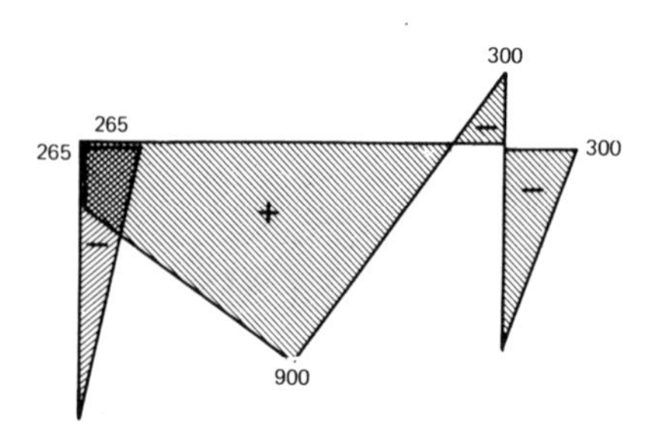

3.5 $\lambda_c = 1.50$

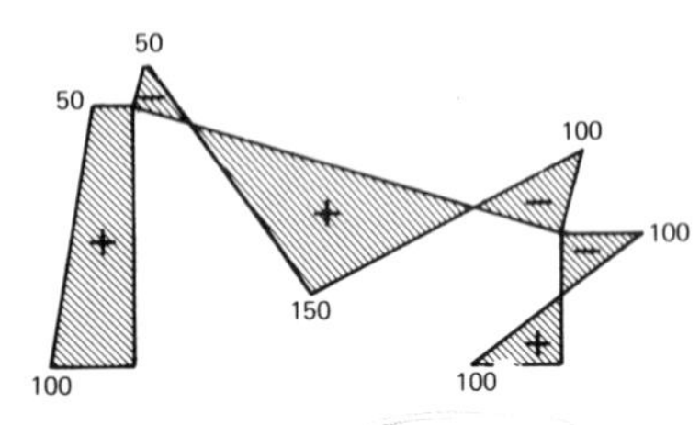

3.6 $V = H$: combined mechanism $V = 2.2M_p/L$; $V = 5H$: pitched portal $V = 4.09M_p/L$

3.7 $k = 0.609$

CHAPTER 4

4.1 (a) $\lambda_c = 1.375$

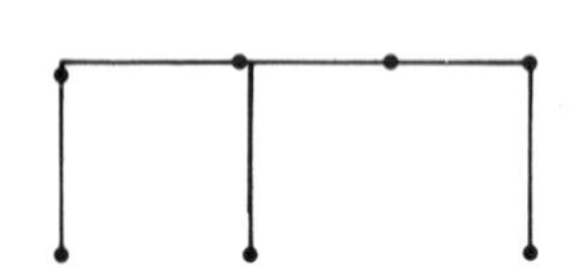

(b) $\lambda_c = 1.448$

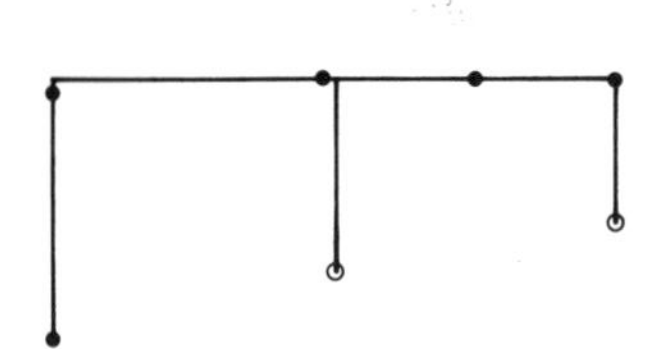

(c) $\lambda_c = 1.80$

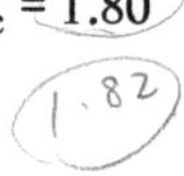

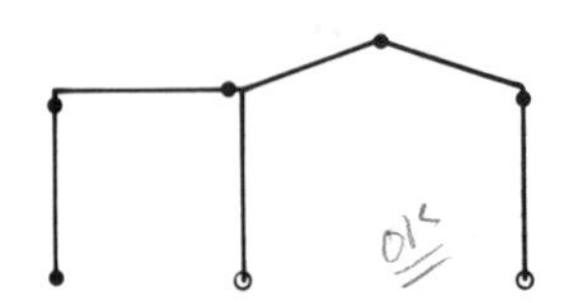

(d) $\lambda_c = 2.00$

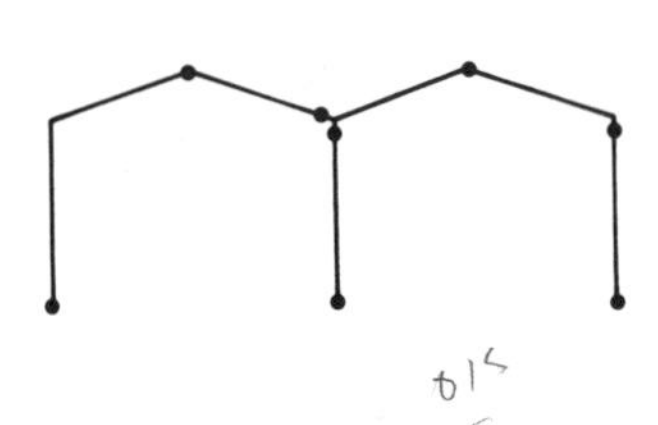

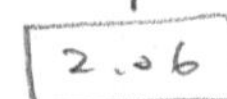

(e) $\lambda_c = 2.235$

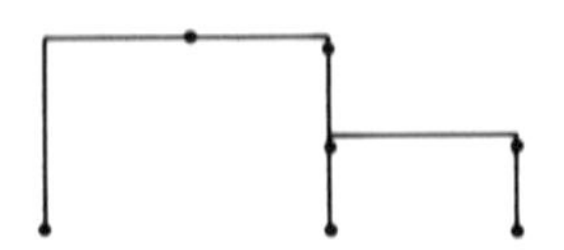

4.2 (a) $\lambda = 1.407$, $1.287 < \lambda_c < 1.407$; (b) $\lambda = 1.191$, $1.169 < \lambda_c < 1.191$

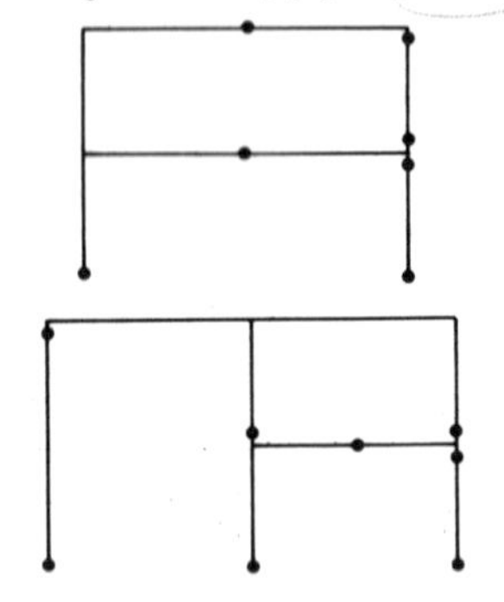

CHAPTER 5

5.1 (a) $M_p = 657.7$ kN m; (b) $M_p = 267$ kN m, plates in BC $M_p = 426$ kN m, plates in CD $M_p = 106$ kN m

5.2 (a) single section $M_p = 2wL^2$ $G = 14.0wL^3$; (b) plated $M_p = 1.13wL^2$, plates in AB and CD. $M_p = 1.31wL^2$, $G = 12.3wL^3$, length $= 1.11L$; (c) minimum weight AB and CD $M_p = 2.44wL^2$, BC $M_p = 1.13wL^2$, $G = 13.1wL^2$

5.3 columns and beams $M_p = 1.167WL$

5.4 $M_p = 0.43WL$

5.5 $M_p = 580$ kN m

5.6 columns $M_p = 133.3$ kN m, beams $M_p = 222.2$ kN m

CHAPTER 6

6.1 $\delta = 2WL^3/243EI$

6.2 $\delta = 0.0077wL^4/EI$

6.3 hinge at A forms last $\delta_h = 0.611WL^3/EI$, $\delta_v = 0.410WL^3/EI$

6.4 hinge at A forms last $\delta_h = 150$ mm

6.5 hinge at ridge forms last $\delta_h = 20.8M_p/EI$, $\delta_v = 10.2M_p/EI$

6.6 at B $\delta_h = 34200/EI$, at C $\delta_v = 23400/EI$, at F $\delta_v = 9000/EI$

6.7 (a) $P = 0.303bd\sigma_y$; (c) intercept of elastic and plastic curves $P_c \approx 0.289bd\sigma_y$; (d) $P_{RM} = 0.275bd\sigma_y$

CHAPTER 7

7.1 $M_r = wL^2/16$

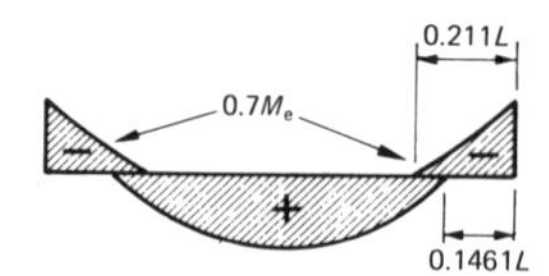

7.2

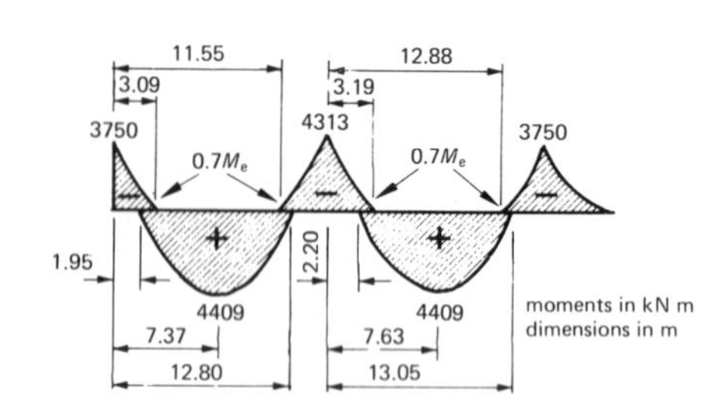

7.3

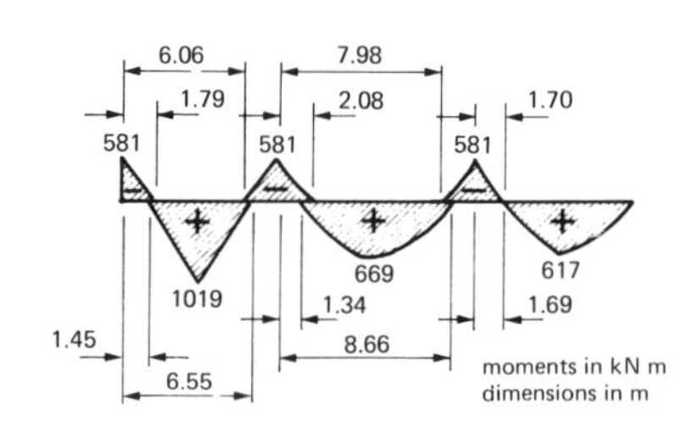

7.4

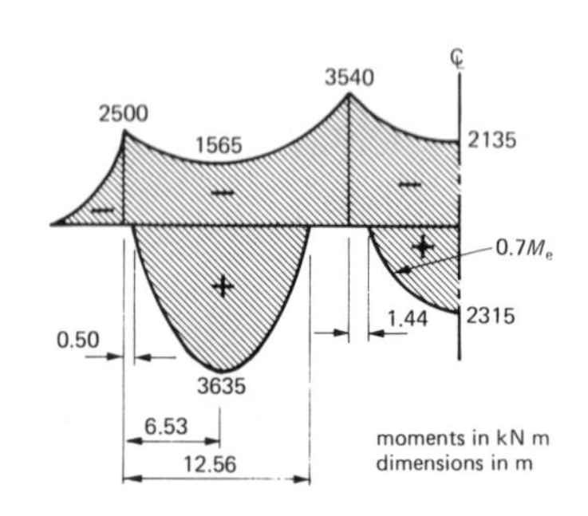

CHAPTER 8

8.1 $q = 17.54$ kN/m^2, in centre of slab $M_r = 8$ kN m/m, at edges $M_r = 12$ kN m/m

8.2 (a) $q = 58.61M/L^2$; (b) $q = 10.7M/L^2$

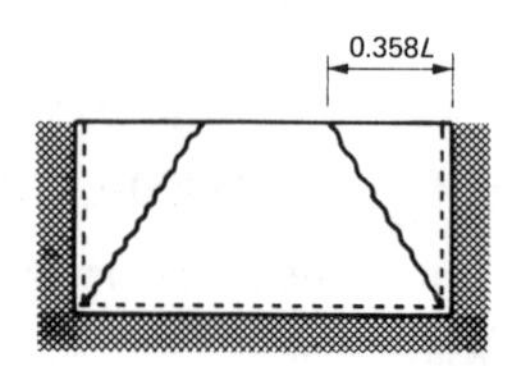

8.3 $M = 7.54$ kN m/m, $x_{crit} = 1.586$ m, $y_{crit} = 2.45$ m

8.4 $M = 0.116qL^2$

8.5 $q_c = 43.85M/L^2$

8.6 $q_c = 24M/L^2$, slab failure; $q_c = 3.25M/L^2$

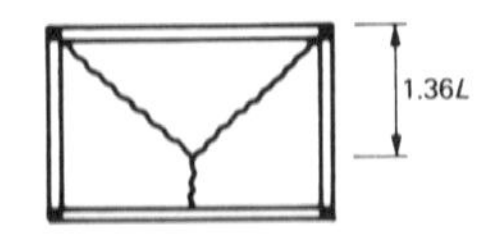

8.7 suggested strip distribution

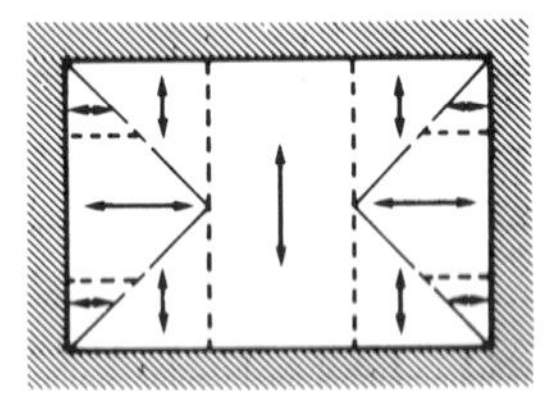

8.8 same strip distribution as in 8.7 but edges are fixed

8.9 suggested strip distribution

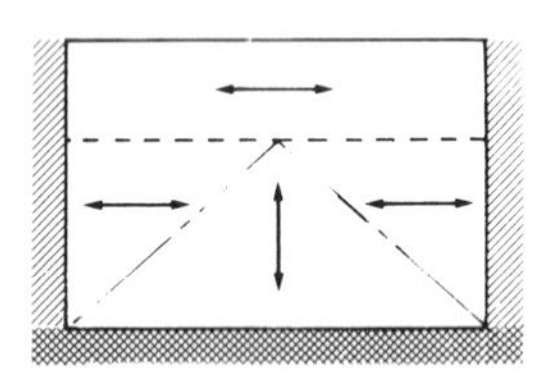

8.10

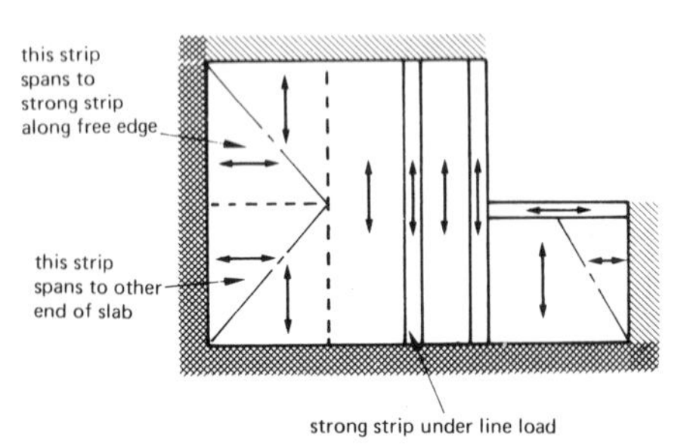

8.11

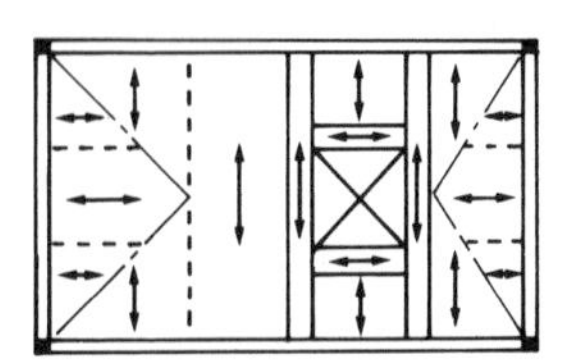

INDEX

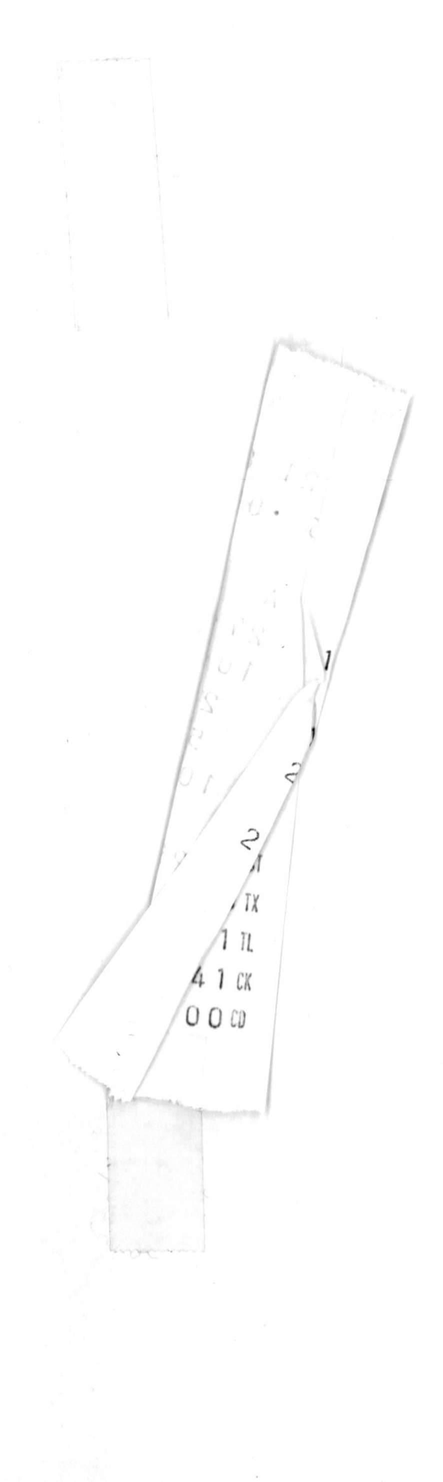